THE
QUIRKS & QUARKS
GUIDE TO SPACE

ALSO BY QUIRKS & QUARKS
The Quirks & Quarks Question Book

THE
QUIRKS & QUARKS
GUIDE TO SPACE

42 QUESTIONS (AND ANSWERS)
ABOUT LIFE, THE UNIVERSE,
AND EVERYTHING

JIM LEBANS

INTRODUCED BY
BOB MCDONALD

CBC radioONE

McCLELLAND & STEWART

LIBRARY AND ARCHIVES CANADA CATALOGUING IN PUBLICATION

Lebans, Jim
The quirks & quarks guide to space : 42 questions (and answers) about life, the universe, and everything / written by Jim Lebans ; introduction by Bob McDonald.

ISBN 978-0-7710-5003-9

1. Astronomy–Miscellanea.   2. Outer space–Miscellanea.
I. Title.   II. Title: Quirks and quarks guide to space.

QB44.3.L42 2008   520   C2007-905844-2

We acknowledge the financial support of the Government of Canada through the Book Publishing Industry Development Program and that of the Government of Ontario through the Ontario Media Development Corporation's Ontario Book Initiative. We further acknowledge the support of the Canada Council for the Arts and the Ontario Arts Council for our publishing program.

Typeset in Palatino by M&S, Toronto
Printed and bound in Canada

This book is printed on acid-free paper that is 100% recycled, ancient-forest friendly (100% post-consumer recycled).

McClelland & Stewart Ltd.
75 Sherbourne Street
Toronto, Ontario
M5A 2P9
www.mcclelland.com

1 2 3 4 5   12 11 10 09 08

"Space is big. Really big. You just won't believe how vastly hugely mind-bogglingly big it is. I mean, you may think it's a long way down the road to the chemist, but that's just peanuts to space."   – Douglas Adams, 1952–2001

CONTENTS

Acknowledgements

To avoid any egregious blunders, we asked some eminent Canadian astronomers (and friends of the program) to look over parts of the manuscript. We'd like to thank them for volunteering their time and expertise. They are Paul Delaney, York University; Brett Gladman, University of British Columbia; Vicki Kaspi, McGill University; Jaymie Matthews, University of British Columbia; Douglas Scott, University of British Columbia; Christine Wilson, McMaster University. Any remaining errors are entirely the fault of the writer.

INTRODUCTION

Let me see if I've got this right: the Universe goes on forever, but we can't see it all at once, because when we look out into space, we look back in time; and when we look all the way back to the beginning of time, we see a Big Bang, out of which the Universe has been expanding at an ever-increasing rate, which will continue until the Big Rip.

Right.

Sound confusing? It is.

After more than two thousand years of studying the night sky, with everything from the human eyeball to powerful instruments that could spot a light bulb on the Moon, the Universe continues to baffle. In fact, even today in the twenty-first century, there is more stuff out there we can't see than what we can. In other words, most of the Universe is still missing.

On the other hand, the only way to solve the mysteries of the cosmos is to ask good questions: "How high is up?" (we don't know); "Is there life out there?" (don't know that either). Okay, so there are more questions than answers, but we've come a long way, especially in the last few decades. And *Quirks & Quarks* has been there – following these developments every step of the way. Most of the topics in this book have been discovered within the lifetime of our program.

If you wanted to see a planet in 1975 (the year *Quirks & Quarks* first hit the airwaves), all you could do was look through a telescope. So planets all looked like fuzzy balls that came in assorted colours. But since then, thanks to a fleet of spindly robots with names like *Ranger, Mariner, Pioneer, Voyager,* and *Pathfinder*, we have reached out to every planet in the Solar System except Pluto – and there's a spacecraft on the way there now. Through their eyes, we have discovered marvellous new horizons with taller mountains, deeper valleys, and wilder weather than anything our little planet has to offer. What were once gods in the night sky have become places that we've flown around, landed upon, and driven across. We've been there and we've been amazed.

The year after *Quirks* went on the air, we got the first colour pictures from the surface of Mars and discovered the Red Planet has a pink sky. It was the first of a long string of surprises that greeted us as we journeyed farther into the unknown. A textbook in 1975 would have told you that Jupiter had twelve moons. Now we've seen at least sixty-one celestial bodies circling the giant planet, in all sizes, shapes, and colours. The moon Io is orange with a dozen volcanoes spewing sulphur crystals at muzzle velocity up into umbrella-shaped plumes that spray the surface with yellow powder. In some places, the ground is so hot the sulphur has melted into black lakes with yellow islands. I'm not making this up.

In contrast, pure-white Europa, as smooth as a billiard ball, is an ice-covered world that may hold a salty ocean with as much water as all the oceans on Earth. Another harbour of life?

Saturn's moon Titan is shrouded in a smoggy nitrogen atmosphere, denser than the air we breathe on Earth. We landed on this moon and found coastlines where methane rain feeds rivers that cut through deep ice valleys and flow into hydrocarbon seas. I'm still not making this up.

We've sent probes through the tails of comets, landed on an asteroid, and produced the first comprehensive atlas of our solar system. We are the modern-day explorers who discovered more than a hundred new worlds just in the last thirty years.

On the other hand, humans haven't ventured more than 400 kilometres into space in the last three decades. That's not very far. But lots of people, including Canadians, have learned to live and work in the strange environment where gravity appears to have taken a vacation. Interesting things happen to the mind and bodily functions when up and down no longer exist. Human spaceflight is currently restricted to the largest construction site ever flown above our heads, the International Space Station, but even after that project is completed, people won't travel very far into space. The Universe is too big and our rockets are painfully slow.

A trip to the Moon takes about two days, which is still manageable. But a return trip to Mars is a three-year journey, while a one-way trip to Pluto takes more than a decade. And forget about reaching the other stars in the galaxy. A journey to even the closest stars would take more than a human lifetime, using current technology. We need warp drive, or something to add a little zip to our spaceships. Of course, we can't go too fast or we start running into all those problems Professor Einstein taught us, about returning home to find your family and friends have aged much faster than you have. Fortunately, we are a long way from having to deal with that problem.

If this journey through the unimaginably big Universe makes you feel small and insignificant, well, you are. Our planet is incredibly puny in the grand scheme of things but it is the only one we know of (so far) where life has evolved. And some of that life has begun to figure out the really big picture. We are fortunate to be living during this amazing time in human history when our eyes are opening to the wonders that surround us.

And that's why we've written this book. My colleague Jim Lebans is an award-winning veteran producer on *Quirks & Quarks* who has followed space discoveries and developments for many years on the program. He wanted the book itself to be a journey through space, from the ground up, beginning with the edge of space just above our heads, right out to the edge of the Universe and the end of time. Along the way, you will pick up some handy travel tips, such as the best way to get from here to there, the best places to visit, and how to make a passing planet more comfortable. You'll also explore some of the biggest questions in science, such as the rubbery, stretchy, dark, time-warped nature of the Universe itself, or whether there's any other life out there. This is a book for anyone who is fascinated by the world around and above them. And as we say on *Quirks & Quarks*, you don't need a Ph.D. to understand or enjoy it.

While we hope the journey through this book will hold many surprises for even the most avid space enthusiast, our journey out into the Solar System and beyond has led to one other big surprise – a new perspective on Earth. Of all the planets we've seen, none of them is like the little blue marble we call home. The biggest lesson we've learned from our travels in space is the value of what we left behind. We live on the crown jewel of the Solar System – not too hot, not too cold, with a breathable atmosphere and liquid water bathing most of the surface. Many space enthusiasts speak of the need to live on other planets in case this one becomes uninhabitable. That isn't going to happen. The planets are very, very far away, difficult to reach and more difficult to survive on. If anything, the planets have shown us how inhospitable the Earth could become if the climate changes in the extreme. Venus is a runaway greenhouse with searing temperatures and sulphuric acid clouds, while Mars is gripped in a permanent ice age where the warmest summer day is about the same as winter in Winnipeg. Jupiter spews out such deadly radiation you can't even get close to it. Saturn is a puffball planet with no solid surface to stand on, and Pluto is so cold that its air freezes solid.

Think of it this way: from what we've seen so far, this blue planet Earth is the only place where you can step outside, take a deep breath, and go, "Ahhh, nice day" – and you're not wearing a spacesuit. Our journey of discovery has really been a discovery of our own unique place in this amazing Universe. I hope this book helps you understand that journey a little better.

Bob McDonald
August 2007
Toronto

WHERE DOES SPACE BEGIN?

THE POINT OF THIS BOOK is to take you on a journey from the Earth out into the Universe to explore space. So the obvious first question is where does space start? Where does that final frontier begin? The quick answer is that the question doesn't make sense. There's no such thing as *not* space – space is everywhere, and even on the surface of the Earth we're in space (we just happen to be in a bit of it that is occupied by a lot of mass). So by this definition, space has no beginning, which isn't helpful to us at all. We want a more interesting answer than that.

So let's go to the aeronautical authorities. According to the World Air Sports Federation (actually, the Fédération Aéronautique Internationale, or FAI), which adjudicates and recognizes all aeronautical and astronautical records, the boundary for space is 100 kilometres. If you can get above 100 kilometres, you're an astronaut. The 100 kilometre limit was arrived at in the mid-1950s by a committee assembled by a Hungarian engineer and physicist named Theodore von Kármán, and so it's called the Kármán line or limit. The logic behind the limit is that at around 100 kilometres you're at an altitude at which you're not really flying any more because the atmosphere is so thin it's aerodynamically insignificant – it's not doing anything useful like providing lift for airplane wings. Since you're not flying any more, then your craft must be operating as a spacecraft rather than an airplane, therefore, you're in space. In other words, the 100 kilometre limit is the somewhat arbitrary point between flying and not flying. (The other point between flying and not flying is the ground, and it's rather less arbitrary.) It is also, of course, a nice round number, which makes it very attractive for record makers and keepers. U.S. military authorities and NASA have set the bar a little lower. U.S. astronauts get their "wings" – a badge for having flown in space – for exceeding 50 miles' altitude – just under 81 kilometres. You could say that in the United States, space is 19 kilometres lower.

Humans got their first experience of space on April 21, 1961, when Yuri Gagarin of the Soviet Union flew *Vostok 1* in a full orbit of the Earth, reaching a height of 315 kilometres, breaking the space barrier rather comfortably. American Alan Shepard was second to the space party two weeks later with a suborbital flight to only 187 kilometres. Marc Garneau became the first Canadian in space when he flew on the Space Shuttle in October 1984, orbiting at 404 kilometres. The title of "first primate in space" went to Albert II, a rhesus monkey who flew to 133 kilometres aboard a modified American V2 rocket in 1948 – more than a decade before Gagarin. Sadly, Albert II was killed when the rocket crashed to Earth. The V2 wasn't designed as a passenger vehicle and had no way to land gently. That said, it wasn't the flight that did him in – it was the sudden stop at the end.

Because it was broken so early and often in the space age, the 100 kilometre limit hasn't had a great deal of significance until recently. In June 2004 *SpaceShipOne*, designed by aircraft innovator Burt Rutan, became the first privately funded craft to make it above the 100 kilometre barrier. The plucky little ship was bolted under the belly of another airplane, hoisted up to about 50,000 feet, and then released so it could light its rocket engine and blast up into space. Three months later it did it twice in six days, and in doing so won the $10-million Ansari X Prize. Now Mr. Rutan's company is planning on offering flights over 100 kilometres to the public in a larger version of the ship sometime after it's built and tested. If you've got $200,000 to spend, it might make a memorable holiday – you'll go into space and you'll get your astronaut's wings.

The problem with the 100 kilometre limit for space is that, strictly speaking, it's still well within Earth's atmosphere. The problem with defining where space begins is that we can't pinpoint where the atmosphere ends. Many satellites in Earth orbit, for example, are essentially still in the atmosphere to some degree, and while you can't use the atmosphere to fly up to orbital heights, it will still bring you down. If you achieve orbital velocity at 100 kilometres, there's more than enough atmosphere to slow you down quite quickly. This is true quite a distance up. The International Space Station (iss), for instance, orbits at about 360 kilometres, where the atmosphere is about a millionth as thick as it is at sea level. The iss, however, is sufficiently slowed by slamming into molecules of air that it drops the equivalent of several kilometres a year. One of the jobs of the Space Shuttle and the Russian spacecraft that visit the station is to boost it back up again. Satellite operators typically consider the effects of atmosphere as far as 1,000 kilometres from the Earth's surface. At that altitude, it will take more than a century for the tenuous atmosphere to pull a satellite out of the sky. So you might say that, for the purposes of a satellite, empty space begins at 1,000 kilometres.

Of course if your definition of space is a complete emptiness – pure vacuum – then it's not just a matter of going farther, as truly empty space may not exist at all. We'll get back to that later.

How do you turn a planet

into a home?

WE'RE PRETTY LUCKY HERE on Earth. We've won the cosmic lottery with odds far worse than you'd ever get in Las Vegas, and the payoff – life on a habitable planet – has been huge. As far as we know (and when it comes to planets other than our own, even those in our solar system, we know very little), Earth is unique. It is the only planet we know of in the Universe on which there is life, and the concatenation of circumstances that led to this is something special.

First we had to be in the right part of the right kind of galaxy, in a pleasant solar system. Even so, making a habitable planet takes some arranging. Scientists have spent a fair bit of time figuring out how our planet became such a delightful place to live. They're satisfying their curiosity in doing so, but understanding how Earth came to be as it is can help guide our search for other Earth-like planets. If we can't find Earth-like planets, this knowledge might help us to understand what planetary engineering would be required to make an uninhabitable world into somewhere we could live. This concept is known as terraforming, and the best example of it we have is the process that nature used to turn Earth from a barren, airless, largely molten rock into the green and verdant land we inhabit today.

The first step is to form a planet. Our solar system – the Sun and all of the planets – coalesced out of a disk of gas and dust more than 4.5 billion years ago. The Sun formed first, then several planets from the remaining material, but only one was habitable. It had to be made of the right stuff and be in the right location. Fortunately these two characteristics go hand in hand to some degree. The planets that formed in the inner part of the Solar System are rocky. Farther out, the planets are gas giants like Jupiter, Saturn, Neptune, and Uranus. The difference between the inner rocky planets and the outer ones is mostly one of size and subtraction.

When the planets formed, they all did so out of a disk orbiting the Sun that was largely homogeneous. It consisted of light gases, chiefly hydrogen and helium, and dust made up of iron, magnesium, silicon, and other heavier elements including, importantly, oxygen, nitrogen, and carbon. The gas and dust coalesced into planets, and the size of the planets is roughly proportional to how much matter they collected in their orbits around the Sun. Jupiter and Saturn are giants because of their distance from the Sun. The farther out a planet forms, the more distance it covers in its orbit, and the more dust it sweeps up. Their size gave them another advantage, as their huge gravity allowed them to hold on to their hydrogen and helium. Jupiter today has a rocky core about fifteen times as big as Earth, surrounded by layers of first metallic, then liquid, then gaseous hydrogen, as pressure decreases farther from the core.

Nearer the Sun, however, the small planets simply weren't big enough to hold on to their gases. Most of the light hydrogen and helium gas was relatively quickly blown away by the solar wind, into the farther reaches of the Solar System. This left the early planets – Mercury, Venus, Earth, and Mars – naked chunks of rock.

The next step for Earth as it underwent a transition from childhood to adulthood, was to re-acquire an atmosphere. Like a volatile adolescent, it did this in an explosive and dramatic way. Early Earth was hot and its interior molten, and for the first couple of hundred million years it spouted volcanoes like pimples on a teenager's cheeks. From these volcanoes the early atmosphere was built, consisting of nitrogen, water vapour, and carbon dioxide. These gases were heavy enough for Earth's gravity to hold on to them. The same thing is likely to have happened on Mars and Venus.

That was the beginning, and so the Solar System had three viable rocky planets (Mercury was too hot to be a possible home for life). Each had roughly the same potential for life and the eventual evolution of sports utility vehicles. So what made the difference?

Like Goldilocks's porridge, it's a matter of temperature, and temperature is a matter of distance from the Sun. Venus was too hot – too close to the Sun. As a result, the water vapour in its atmosphere never rained out. When water rains out of the atmosphere it takes some carbon dioxide with it, which then returns to its mineral form in the ocean. Since this didn't happen on Venus, the greenhouse effect, which is fuelled by water vapour and CO_2, ran away with itself. So Venus became a baking furnace. Its atmospheric temperatures were measured in 1994 by the *Magellan* spacecraft, which orbited the planet many times before plunging down through the Venusian atmosphere. The probe showed that at the surface it was hot enough to boil lead. Needless to say, the *Magellan* probe didn't last long in the Venusian atmosphere.

Mars, on the other hand, is too cold. While its early atmosphere might have produced a greenhouse effect similar to Earth's, it wasn't able to sustain it. This might have been because asteroid impacts blew away its atmosphere and it didn't have enough volcanoes to replenish its CO_2. As a result, Mars froze. It now has a wispy atmosphere far too thin to support life.

Earth, of course, was perfect. Its orbit was at the right distance from the Sun for the optimum amount of solar heat (though a little closer or a little farther away would probably have been fine as well). Its size was just right for continuing tectonic activity, which recycled the land and drove the volcanoes that contributed to our atmosphere. We had a few more lucky breaks as well, including a magnetosphere – a magnetic shield that deflects radiation from space. The magnetosphere is thought to be generated by Earth's partially molten, rotating iron core. Other planets either lack such a core or theirs has frozen and no longer generates a magnetic field.

It took a few hundreds of millions of years for early unicellular life to develop and then transform the atmosphere of the planet by converting the CO_2/nitrogen atmosphere into the oxygen/nitrogen atmosphere necessary for complex life forms.

We can also thank our giant planetary neighbours – particularly Jupiter – for intercepting most of the comets that occasionally sweep in from the outer reaches of the Solar System. This makes events like that unfortunate collision that killed off the dinosaurs rare, allowing the important evolutionary business on our planet to go on with only occasional interruptions for many hundreds of millions of years.

And unlike poor Goldilocks, we kept the bears under control.

What cosmic catastrophes could wipe out all life on Earth?

As if we didn't have enough to worry about, it's certain that we're headed toward a mass extinction that could wipe out much or all life on Earth. It has happened before and will happen again. In some cases there may be something we can do to prevent a catastrophe. In others there's nothing at all we can do, and what's more we might not even see our doom coming. Cosmic catastrophes that could wipe us out include asteroid impacts, nearby supernovae, gamma ray bursts, and rogue black holes. It's a terrifying list.

Extinction events are a big part of Earth's history. The fossil record shows that much of the life on Earth has been wiped out several times already, and it appears to take the survivors tens of millions of years to repopulate the planet. The most famous of them occurred 65 million years ago and wiped out the dinosaurs, but they were far from the only casualties. During this calamity, which is called the Cretaceous-Tertiary extinction (or the KT extinction), we lost as much as half of all the species on the planet. The number of species lost, however, is a deceptive measure of the destruction. The most popular theory for the cause of this extinction was a 10 kilometre wide asteroid smashing into the vicinity of what is now the Yucatan in Mexico. In 2007, astronomers determined that this asteroid was probably deflected toward Earth by a collision between two rocky bodies 160 million years ago in the asteroid belt. The impact of this cosmic pool ball would have annihilated everything within hundreds or thousands of kilometres. It would have created a global firestorm, setting most of the planet's forests on fire. This devastation was likely followed by a long "nuclear winter" caused by the vast amount of dust, debris, and smoke thrown up into the atmosphere blocking out the Sun. The catastrophe must have been unimaginable, and the way paleontologists usually talk about it minimizes the scale of the disaster. They point to the number of species that went extinct leaving no descendants at all. As we mentioned, that was roughly half of the species on Earth at the time. The actual loss of life would have been much greater because, even if 99 per cent of a species were killed, the species might still survive to breed and repopulate the planet. It's entirely possible the disaster killed upward of 90 per cent of life on the planet.

The KT extinction wasn't even the largest extinction event. It may be third or fourth on a list that most likely includes more than a dozen. The biggest was the Permian extinction, about 250 million years ago, in which upward of 90 per cent of the species in the oceans and 70 per cent of the species on land disappeared. The terrifying thing about this mass extinction, and

the other dozen or so events in which life was nearly extermi-
nated, is that we don't really know what caused them. There are
several theories concerning each extinction, and some of them
include disasters that could happen again at any time.

Asteroid or comet impacts are one of the big ones, and we're
certainly not safe from those in the future. We're confident that it
was an impact that wiped out the dinosaurs, and some scientists
have suggested that other extinctions, including the big Permian
event, were also caused by catastrophic impacts. Over the history
of the Solar System, impacts have grown gradually less likely as
many of the objects floating around in space have already hit
something. Jupiter, in particular, acts like a vacuum, sucking in
stray objects that drift into its colossal gravitational field. Despite
this, it's been calculated that an object bigger than 2 kilometres is
likely to strike the Earth about every 2 million years. That would
cause a global catastrophe but it might not be violent enough to
cause massive extinctions. It could certainly be a large enough
disaster to kill off a good part of the human population and
threaten our civilization. Bigger objects may well be out there
too, but it's harder to calculate the odds of those kind of impacts.
They will definitely be less frequent.

Currently, small-scale sky-survey programs are trying to
identify and track the objects that might pose a risk, and
astronomers think they've identified about two-thirds of the
large ones. So far they haven't found one that's going to hit any
time soon. This is not to say we haven't had close calls. We've
tracked asteroids drifting by us in space at distances far closer
than the Moon is to us – a hair's breadth in space measurements.
It's also entirely possible that an asteroid that hasn't been
spotted yet, or even a rogue object like a comet coming in from
farther out in the Solar System where we can't track things, is on

a collision course with us. There's no guarantee that we would see it before it hits. The first indication we might have of it is the flash of impact or the earthquake that would result from its smashing into our planet.

We *may* be able to do something about an asteroid coming at us if we see it soon enough. Hollywood disaster movies have shown us brave astronauts using atomic weapons to shatter a large asteroid only days before it's likely to hit us. This probably would not work and might well make things worse as the broken hunks rain down on us. Our best chance of avoiding a collision is to spot an oncoming asteroid years or decades before it's likely to hit us, and then use a spacecraft to gradually push it into another orbit, just enough to miss us. The earlier this is done, the less push is needed, which makes the operation much more practical than blowing it to smithereens.

Asteroid impacts are the mostly likely disaster we're going to face and at least there's a chance we could do something about them. We might not be able to avoid other kinds of cosmic catastrophes, however. One possibility is that a star in our galactic neighbourhood explodes, bathing us in a blast wave of radiation that wipes out all life on the planet.

An exploding star is called a supernova, and it occurs when a large star burns out all of its nuclear fuel. A star is kept in a delicate balance between gravity, which wants to pull material in and make it denser, and the outward pressure exerted by the continuous thermonuclear fusion reaction in its cores. When the fuel at the core is exhausted, the star collapses in on itself. The huge mass of the star is compressed into a tiny space and then recoils outward in a titanic explosion that blows the star apart. A vast amount of radiation is released, including visible light, ultraviolet light, and gamma rays, but also high-energy atomic particles like protons, which tear through space at close to light speed. If the radiation of a nearby supernova was to hit Earth, it would be a disaster, shredding our magnetic field and bathing the planet in radiation that would kill all life not buried deep in the planet. A supernova occurring a little farther away could still produce enough radiation to destroy the ozone layer in our upper atmosphere. Ultraviolet radiation from the Sun would stream in, destroying plants, which would result in the collapse of the food chain. There would be widespread starvation across the animal kingdom. We can't be sure, but supernova explosions like this are the suspected culprits in some of Earth's earlier major extinctions.

Fortunately this scenario turns out to be less likely than we'd previously thought. Calculations made in the 1970s suggested a supernova within 50 light-years would be close enough to extinguish life on the planet. More recent work suggests that the explosion would have to be much closer than that – not more than 25 light-years away. Happily for us, there don't seem to be any stars in our neighbourhood large enough to be a supernova candidate. There is one star about 150 light-years away that's been identified as a candidate, but it's not likely to blow for at least 100 million years and it's at a safe distance. Stars do drift around in our galaxy, though, and a supernova candidate could be headed toward us. A star cloud called Sco-Cen, for example, which contains several short-lived, massive stars – perfect candidates for supernovae – drifted to within 130 light-years of us about 5 million years ago. Had it come closer and one of its stars exploded, it might easily have destroyed our hominid ancestors before they got out of Africa.

A supernova much farther away could still destroy life on Earth, but it would have to be a special kind of stellar explosion that produces something called a gamma ray burst. Gamma ray bursts are thought to be the result of particularly large stars exploding and then collapsing into a black hole. They may also result from the collision of two ultra-dense neutron stars. In either case, they send a powerful directional beam of radiation across space, lasting perhaps only about ten seconds. These are rare events, and we've never seen a gamma ray burst in our own galaxy, but they've been observed coming from other galaxies. Our galaxy is thought to have the kind of supernova that could result in a gamma ray burst about every 500 years, but 99 per cent of the time the beam would not be pointed in the direction of Earth. If one does occur in our galaxy and it is pointed at Earth, we should be safe unless it happens relatively close – within about 3,000 light-years (the centre of the galaxy is about 30,000 light-years from us). It's been calculated that this happens roughly about every 100 million years. Like a nearby supernova, gamma ray radiation would strip away the ozone layer, and planetary devastation would follow. It's entirely possible that, in the past, a gamma ray burst has affected the evolution of life on Earth.

One final disaster that would end it all for us is a close encounter with a rogue dwarf star or small black hole. Most of the stars in our galactic neighbourhood orbit with us around the centre of the galaxy in a relatively well behaved manner. It is possible, however, that a wandering stranger could come our way, and this could be very bad news for our planet. It might be a brown dwarf star, one that is too small to maintain fusion and so becomes dim and difficult to detect. A small black hole could also drift through the galaxy, essentially invisible to telescopes. If one of these drifted through our solar system, its gravity could tug us out of our orbit, dragging us into a more distant and colder orbit from the Sun, or pushing Earth into a closer and hotter one. Either way, it would be a disaster. Earth could even be flung out of the Solar System entirely. Even without a close encounter, one of these massive dark bodies drifting through the outer reaches of the Solar System could disrupt the regular orbits of the comets and planetesimals drifting out there, throwing them toward us and vastly increasing the chances of a killer impact with Earth. We actually have no idea how likely an event like this is. We haven't seen any that would pose this kind of threat, but brown dwarfs and small black holes can be quite difficult to detect. The truth is that we don't know how common they are, or how many might be in our galactic neighbourhood.

Humans haven't been around very long on this planet, and we shouldn't waste time feeling anxious about these major extinction events, not least because there's not much we can do about them. We might even want to celebrate them. After all, if that asteroid hadn't wiped out the dinosaurs, giving mammals the opportunity to evolve and diversify, we would not be here right now. Or perhaps we would but we'd all be speaking Velociraptor.

WHY ARE ROCKETS A LOUSY WAY
TO GET INTO SPACE?

HERE'S A LITTLE ANALOGY FOR one of the things wrong with our attempts to explore space: the first tools humans used were made out of rocks and pieces of wood and probably sinews. Today, our tools are made out of sophisticated metal alloys and synthetic composites. Modern tools are better. When humans first started going into space, our propulsion systems were chemical rockets. Now after decades of advances in space travel our space vehicles are powered by, you've got it, chemical rockets. In fact, the tool analogy can be stretched further. One of our species' great early technological breakthroughs occurred in ancient prehistory when we harnessed fire. Today our most sophisticated space vehicles are still powered by fire – the combustion of chemical fuel. This is what's really holding us back from taking the next step into space.

Of course, we've increased the efficiency and reliability of chemical rockets considerably since the first flights into orbit in the 1950s. The Space Shuttle's main engines are among the most efficient rockets ever built. They can generate immense amounts of power, using nearly every bit of energy in their fuel, but they consume vast amounts of fuel. More than 90 per cent of the 2 million kilogram launch weight of the shuttle is fuel, and most of it is burned in the first two minutes of flight. The energy that represents is huge. The fuel for a loaded Space Shuttle has the energy equivalent of about 4 kilotons of TNT – about 20 per cent of the energy in the atomic bomb that destroyed Hiroshima.

Yet all this is barely sufficient to get the shuttle a couple of hundred kilometres off the ground and into low orbit. The Saturn V rocket that sent the Apollo missions to the Moon was even more profligate in its energy use. What's achieved for all this? The shuttle's payload is about 25,000 kilograms. So that means 80 kilograms of fuel are used to put 1 kilogram of payload into orbit. At a launch cost of about $500 million, that means that each kilogram in orbit costs $20,000, just for shipping. The price for an astronaut, and we'll assume she or he averages a fairly trim 65 kilograms, would be about $1.3 million – and that's not including food, water, oxygen, and clothing. Thanks to gravity, the return trip is pretty much free, but you have to admit, it's still a pretty pricey ticket.

NASA has acknowledged that the Space Shuttle has to be retired and is designing a new vehicle to replace it in the job of carrying humans and cargo into space. In many ways, however, the new vehicle will not be an advance, but a kind of step backward. Early mock-ups of the new Crew Exploration Vehicle, which is meant to fly by 2014, look very much like super-sized versions of the Apollo capsule that returned astronauts from the Moon in the 1970s. It will ride on the top of a rocket that has lower stages derived from the design of the Space Shuttle's solid rocket boosters – the "candlesticks" that ride on the side of the shuttle's main fuel tank – and an upper stage powered by an updated version of the Apollo rocket engine. This may turn out to be a more efficient way to get humans into space than the shuttle. It may take humans to the Moon and possibly Mars. It doesn't, however, seem like a great leap forward.

Private industry is trying to make rocket travel less expensive and more efficient by cutting out NASA bureaucracy and overhead. An assortment of space enthusiasts and dot-com billionaires have poured millions into developing private launch systems. Aerospace maverick and pioneer Burt Rutan's *SpaceShipOne* demonstrated in 2004 with the first private manned flight into space that it can be done. Rutan's group hopes to develop its vehicle into a high-priced thrill ride to take tourists into space for a few tens of thousands of dollars. His spacecraft, however, doesn't achieve anything like the tremendous speeds necessary for orbit. It can get into space barely and briefly, but it can't stay there. To achieve orbit it would have to be much larger – and more expensive. The private space launch venture SpaceX is more ambitious and leads private industry in developing satellite and manned orbital launch vehicles. The most ambitious crafts these private efforts build might reduce the cost of getting passengers and material into orbit by a factor of ten. This would bring the cost of space launch down from astronomical (excuse the pun) to just fantastic.

Unfortunately the problem with rockets is not going to be conquered with better design and more exotic fuels. Rockets use pretty much the most energy-rich sources of chemical energy available, and tweaking the way they work is simply not going to radically alter their efficiency and cost. The real problem is intrinsic to their design. Chemical rockets use a lot of fuel, and they have to carry that fuel with them. If you think about it, most of the fuel in a rocket is used to boost fuel upward. It's only at the very end of its flight that the Space Shuttle outweighs the fuel it's carrying. In fact, the Space Shuttle could never get into space if it wasn't constantly getting lighter by burning off fuel. It's riding a vast pyramid of energy to get into orbit.

Rockets can, just barely, do the job of getting us into space, but the cost in money, in fuel, and also in environmental impact makes it impractical to think that we'll ever use them on the kind of scale that would make travel into space routine. What's needed for that are brand-new technologies.

HOW SHOULD WE
GET INTO SPACE?

IF WE'D LIKE TRAVEL TO SPACE to be as easy and inexpensive as airline travel, we'll need new technologies that are more advanced than riding a chemical explosion to climb out of the atmosphere. Fortunately a whole host of methods have been proposed to get us into space more easily. Some are clearly futuristic, demanding technologies and materials that are only theoretically possible. Some, though, have more immediate potential.

The first step, and one that has many enthusiastic proponents, is to give traditional rockets a bit of a boost. This is how the first airplane flew into space. In 1963 – two years after Yuri Gagarin became the first person in space – American Joe Walker flew the X-15 rocket plane above 100 kilometres, reaching speeds of nearly 6,000 kilometres an hour. He thus became the first person to "fly" into space, since his craft actually had wings and used aerodynamic lift (as well as rocket propulsion) to fly out of the atmosphere, rather than just the brute rocket power of Gagarin's modified missile. Another secret of the X-15, though, was that it had a boost. It was flown up to about 45,000 feet, or the first 13 kilometres of its climb, attached under the wing of a B-52 bomber. It was then released from the bomber, fired its rockets, and was off to space.

The benefit of the boost is that flying even to the relatively low altitude of 40,000 feet (at a speed of a few hundred kilometres per hour) gives all sorts of advantages to a rocket. A vast amount of energy is used by rockets getting to that altitude. The Space Shuttle, for example, needs all the power of its main engines and solid rocket boosters just to get to 33,000 feet (and a speed of about Mach 1.3). At that altitude it drops the boosters. Avoiding the effort of that early climb by hitching a ride on a much more fuel-efficient airplane is one way to make a rocket launch cheaper. It's also a proven method. Since 1990, the Pegasus launch system developed by Orbital Sciences Corporation has been doing this to get small satellites into low Earth orbit. *SpaceShipOne*, the first private human mission into space, in 2004 used the same technique. The ship rode beneath a special jet carrier called *White Knight* up to 40,000 feet before separating and lighting its rocket engine for the rest of the climb. The system doesn't even have to use aircraft. The U.S. Air Force has experimented with balloons that can carry rockets to even greater heights. Several private space projects have proposed using a balloon or an airship as a "first stage" for a rocket.

The limit of the boost is that, so far, it's been used only for relatively small vehicles going to relatively low orbits. It also would take a very large airplane to lift a substantial spacecraft and its fuel for a trip into space. A Boeing 747, for example, can carry the empty Space Shuttle – that's how the shuttle is ferried back to the Kennedy Space Center in Florida when weather forces it to land in California. It can't, however, carry a shuttle, its full load of cargo, and its fuel. It is simply too heavy.

An alternative booster might be a high-speed airplane that uses an experimental jet engine to achieve spectacular speeds. It's called a scramjet, and it's optimized for high altitudes, thin air, and very high speeds. The advantage of the scramjet over a rocket is that it needs to carry much less propellant than a rocket does. Rockets, by design, carry both fuel and the oxygen necessary to burn it. The shuttle, for example, uses liquid hydrogen as a fuel, but it needs oxygen to burn the hydrogen and has to carry the oxygen. This is a major drawback because the oxygen weighs three times the hydrogen fuel it burns. A scramjet, on the other hand, harvests oxygen from the atmosphere, so it doesn't pay the massive weight penalty of rocket fuel. NASA's X-43A scramjet test vehicle has now reached Mach 10, or more than 12,000 kilometres an hour, at heights of 100,000 metres. Potentially a scramjet might be able to go higher and achieve speeds of more than Mach 15, over 18,000 kilometres an hour. This still isn't fast enough to go into orbit, but if a scramjet boosted a rocket, the rocket would need far less fuel for the last leap.

The real trick would be to get beyond rockets, even for a last stage. This isn't a new idea. It was first considered in 1866 by Jules Verne in his book *From the Earth to the Moon*. His space vehicle was essentially a bullet fired into space by a gun barrel three football fields long, using a charge of 120 tons of explosives. The surprising thing about this idea is that, while it wouldn't have worked, it did have some technical merit. Canadian scientist Gerald Bull put the idea into practice in 1963. Funded by the United States and Canada, he shot a specially designed 16-inch projectile above 100 kilometres – in other words, into space. Later in his life, Bull, who went on to an unsavoury career as a weapons designer, may have designed and begun to build a much larger supergun, capable of putting shells not just into space, but into orbit. Unfortunately he was thought to be building this remarkable gun for Saddam Hussein's Iraqi government, and was assassinated in Belgium in 1990 before he could complete the project.

The advantage of the supergun is that it's designed so that no fuel need be lifted off the ground. The energy to carry the payload into space all comes at the launch, as it's accelerated very quickly up to orbital velocity and then just coasts into space. All of the energy in the launch goes into lifting the payload rather than lifting the fuel. A supergun driven by a chemical explosion isn't the only way this concept could be used. A similar concept that many scientists think has great potential is to use magnetism in something called a "mass driver." A mass driver is, basically, a giant magnetic catapult. It uses powerful electromagnets to accelerate a capsule along a track perhaps five or ten kilometres long. By the time the capsule leaves the track, it would be travelling at more than 27,000 kilometres an hour, and would shoot up through the sky into orbit. This is the same principle behind particle accelerators used in physics experiments – except the particle would be a spacecraft instead of a few atoms. The problem with mass drivers and superguns is that both systems would subject passengers to crushing acceleration – probably in excess of two thousand times the force of gravity. While this rules them out for carrying humans into space, it doesn't make them impractical for more robust kinds of cargo that humans would need in space, including food, water, fuel, and air. This could be the cargo launch system while humans use another method to get their fragile bodies into space.

That method might be riding a beam of light into the sky. The idea of using a laser to push a spacecraft into the sky has been demonstrated with a vehicle called the Lightcraft, designed by Leik Myrabo of the Rensselaer Polytechnic Institute in New York, working with the U.S. Air Force. His system works by focusing pulses of laser light on a curved mirror at the bottom of the spacecraft. The light energy is concentrated by the mirror and instantly heats up air immediately below the craft to five times the temperature of the Sun – causing it to explode. The force of this pushes the Lightcraft higher. The Lightcraft, which is currently about the size of a hubcap, has flown to more than 40 metres in height.

Forty metres doesn't seem all that impressive, but rocket pioneer Robert Goddard's first liquid-fuelled rocket rose only 13 metres on its first flight. Forty years later humans rode the same kind of rocket into orbit.

WHY NOT JUST TAKE
AN ELEVATOR TO SPACE?

BEFORE EXPLORING AND SETTLING SPACE can become a reality, we need to develop an efficient and cheap way to get there, and the key to that is altitude. There is a mundane technology we already use every day that might do the trick. It's cheap and effective, and it might just be able to get us into space. It is, in fact, nothing more than an elevator.

Elevators have been around for more than a hundred years, and they were one of our first solutions to the problem of gaining altitude short of climbing a mountain. Before the elevator, most people never went higher than four or five storeys, simply because it was too many stairs to climb. With the elevator, we started to scrape the sky. Now you might reasonably suggest that as a method of travel into space, elevators seem to have a critical weakness. They don't really go all that high – far less than a kilometre in even the tallest buildings – and of course you need to have a building to house the elevator, and we're not building a skyscraper that will reach into space any time soon.

The idea of a space elevator has been around since the 1960s when Russian and American engineers independently developed the idea of a satellite tethered by an incredibly long, super-strong cable to a station on the ground. They suggested that vehicles could travel to and from the satellite by scaling and descending the cable. To make the system work, the satellite had to be in geo-stationary orbit, about 36,000 kilometres up, directly over the equator. At this distance the speed that a satellite needs to be going to stay in orbit corresponds exactly to the speed of Earth's rotation, so the satellite, which is orbiting Earth, appears to stay fixed in the sky above it. Geostationary satellites are very useful, of course, especially for communications – things like phones and television. Connecting them to Earth with a cable, however, is not something we've yet attempted.

The first difficulty is how to put the cable where it needs to be – connecting it between the satellite in orbit and the ground. It isn't possible to trail a cable behind like a fishing line as the satel-lite is launched into orbit, as it would slow or misdirect the ascent. So the cable would have to be lowered from space. The lowering is a tricky and complex task. If you lower a heavy cable from an orbiting satellite, you're changing the centre of gravity of the satellite, and it has to adjust its orbit to compensate. In order to obey the law of conservation of momentum, the satellite has to respond by moving upward to keep its centre of gravity in the same place – in geostationary orbit. What this means is that a cable long enough to reach Earth from geostationary orbit would be too short. It would have to be much longer because the more cable that is lowered, the higher the satellite rises, the more cable is needed to reach the ground. The best design seems to involve a cable more than 90,000 kilometres long – long enough to wrap around the world more than twice. Needless to say, we've never made a cable anything like this.

The cable would also have to be incredibly strong – strong enough to support its own weight and the weight of the loads it would carry as vehicles moved along it. In the early studies of the space elevator, finding a material strong enough to make the cable seemed to be a deal killer. Metals like steel aren't anywhere near strong enough and are far too heavy. Advanced synthetics couldn't cut it either. Even super strong diamond fibre wouldn't have the tensile strength to do the job. Without a sufficiently strong material for the cable, the space elevator was going nowhere. Work on the idea stalled.

Then, in the early 1990s, the space elevator concept was given new life. A new material was developed that might be strong enough to be used in the elevator cable: carbon nanotubes. A nanotube is a tube-shaped molecule consisting of carbon atoms joined in a strict geometric form. The strength of the form and the molecular bonds gives carbon nanotubes a tensile strength thirty times that of the strongest steel. What's more, they are extremely light. Spurred by the discovery of this new material, a small group of researchers funded by NASA's Institute for Advanced Concepts began to develop the idea of the space elevator again. They produced a detailed and theoretically workable design.

A carbon nanotube cable could be carried into orbit in pieces. A couple of shuttle missions would bring up the parts for the elevator, and it could be assembled in low-Earth orbit. It could then be boosted up to geostationary orbit, where it would start to unwind its cable.

The cable itself would be a flat sheet of nanotube fibres. Among the most remarkable things about it is how small it would have to be. The design proposed to NASA was for a cable only 10 centimetres wide and a single micron thick. That's about a hundred times thinner than a sheet of paper, and remember, this cable is going to have to be 90,000 kilometres long. The entire cable would weigh on the order of 20,000 kilograms, which means that each kilometre of cable would weigh only a little over 200 grams.

Once this initial cable is strung, a small robotic climber could ascend it, bringing up a second cable behind it. This would be bonded to the first cable with a strong epoxy glue, making it stronger. More cables would be added, until it was strong enough to carry bigger climbers. These would be the real workhorses of the elevator, carrying a 20,000 kilogram payload on each trip. This is about the same cargo capacity of the Space Shuttle.

The elevator cars would be a special design, somewhat different from those in a skyscraper, which are pulled upward by a cable. On the space elevator, the cars or "climbers" would pinch the cable between rollers powered by electric motors and climb right up. The cars would get their power from light-absorbing panels, similar to solar panels, that are fed power by ground-based lasers pumping energy into them as they ascend into the sky. These sorts of systems have already been developed and tested by students in university competitions for prize money supplied by NASA, and they work. Of course, as yet they'd only ascended cables of a few hundred metres in height, but the principle has been proved.

The cable will have to be anchored at the ground, most likely to a floating platform similar to an offshore oil-drilling platform. This has several advantages over a land-based platform, chief among them mobility. The platform would be able to move around, dodging bad weather and also swinging the cable out of the way of orbiting hazards like abandoned satellites and other space junk.

If all this works, we'll have a remarkable tool for getting people and cargo into space at a fraction of the cost of a conventional rocket. Despite the fact that its upper terminal is stationary, a space elevator could be far more versatile than the Space Shuttle for space launch. Passengers could ascend the cable in comfortable pressurized capsules. It could carry satellites up far higher than the Space Shuttle can go and then release them. A small rocket could then be fired to move the satellites to the preferred orbital location. A vehicle could even be lifted up the cable far past geostationary orbit – out to perhaps the end of the cable. From there, just releasing it would slingshot it out into space, to the Moon, Mars, or farther. In effect, the space elevator would double as a space launch system. It would revolutionize space travel, exploration, and exploitation.

That's the dream. Now the reality check: one obvious problem is that a 90,000 kilometre cable thinner than a sheet of paper is going to be vulnerable to all sorts of potential damage and would be difficult, if not impossible, to repair. Lightning strikes could vaporize it. Meteorites or space junk could punch holes in it. Hurricane-force winds could, conceivably, even break it. High in the atmosphere it could be attacked by atomic oxygen, which could chemically erode it away. All these are problems that do not yet have good solutions.

An even bigger problem is that we don't know how to fabricate such a cable. The longest nanotube fibres made so far are tiny threads only centimetres long, and a space elevator cable would be hundreds of millions of times longer, more than 40,000 kilometres. Of course, only a decade ago the longest nanotubes were less than a micron in length, so we've already improved by a factor of 200,000. We are making progress.

Perhaps the biggest problem is that the elevator would be very hard to test in increments. When we started testing rockets, we went a little at a time. First suborbital, then orbital, then getting bigger and more capable every time until we could put together a Moon mission or a space shuttle. But we can't build a short elevator, and then a longer and larger one. We can build and test components individually, but eventually we would have to go big or go home. The first space elevator would have to be a full-scale project certainly costing upward of ten billion dollars. It may have a huge payoff, but it would still be a huge gamble.

Can I run fast and jump into orbit?

The short answer is yes, under certain conditions. It's all a matter of speed and, to some extent, technique.

For inspiration, let's turn to one of our favourite books, Douglas Adams's *Hitchhiker's Guide to the Galaxy*, which gives this technique for flying: "The knack lies in learning how to throw yourself at the ground and miss. . . . Most people fail to miss the ground, and if they are really trying properly, the likelihood is that they will fail to miss it fairly hard."

Whether this is the recipe for flying, we'll leave to your own spirit of inquiry and liability insurance. But it is a pretty good recipe for orbiting Earth. Orbiting, in fact, is simply circling an object at a speed that's high enough that as you fall, you miss. The trick, of course, is getting that speed just right, otherwise you'd fail to miss – fairly hard.

A simple orbit is a balance of speed and gravity. You can see this when you swing a rock at the end of a string, with the string playing the part of gravity. Swing hard enough and the rock will describe a circle – an orbit of your hand. However, if you don't give it enough speed, the rock will fall out of orbit. You see the same thing with rockets. A rocket that goes straight up from Earth's surface never goes into orbit. It can reach the height of something in orbit, like the International Space Station (ISS), which orbits at about 360 kilometres, but the ISS would whiz right by the rocket since it goes around the planet at a speed of about 27,000 kilometres an hour. As soon as the rocket stops accelerating, assuming it hasn't reached a speed that will take it entirely beyond Earth's gravitational field, it will start to drop.

To get to orbit, the trick is to gain enough speed not only by accelerating upward, but by travelling over Earth's surface. In space jargon, this is speed "downrange," and you might have noticed it if you ever saw a spacecraft launch. The rockets rise initially, but they then turn off the vertical to the horizontal, gaining altitude slowly, but swiftly increasing their speed across Earth's surface. The trick for arriving in orbit is to continue to increase that speed to the point that, when you stop accelerating, you're going fast enough to fall around Earth, instead of onto it. In other words, you fall to Earth but miss. What you've accomplished, in fact, is a careful balance of the inertia of your speed over the Earth, and the force of gravity pulling you down.

This is a slightly complicated preamble to the real answer to the question. This might be surprising, but it is at least theoretically possible to run into orbit. That's "theoretically" in the sense that if we ignore enough reality it just might happen. Certain special conditions must apply. To begin with, you'll have to run faster than any human can. We'll also have to make some minor alterations to our planet. One problem is the atmosphere, which makes orbiting considerably more difficult, especially at low altitudes, as it acts as a constant brake on speed. Friction with the atmosphere is why cars, bicycles, and rockets can't go faster than they do, and it's also why meteors (and space junk) don't hit Earth at unimaginable speeds, making huge craters. They're slowed down, and heated up, by friction with the atmosphere. This friction is significant even when the atmosphere is at its thinnest. At the boundary to space (see Question 1), 100 kilometres high, the atmosphere is still so thick it prevents any object from going fast enough to enter a sustainable orbit. There's just too much friction with the air to maintain speed. It's difficult to achieve any kind of stable orbit below about 185 kilometres, and even at this height orbits quickly decay as the spacecraft is slowed by slamming into molecules of air, and then it falls back down. So to run into orbit we need to lose the atmosphere.

Now we've got to do something about the planet itself. Earth isn't round. It's quite lumpy, and it bulges at the equator. That's going to complicate things. It's also covered in water, which you can't run on. So let's transform Earth into a featureless solid sphere. That was easy. Now start running.

You'll have to run pretty fast, at least at 28,500 kilometres an hour (a number you can calculate easily, given a simple formula for orbital velocity). Get to that speed, and then just jump up slightly, so that you're no longer in contact with the ground – even inches will do. You will now be in orbit, travelling completely around the world in a little less than an hour and a half. With no atmosphere to slow you down, and no mountain ranges to run into on this perfectly spherical (and sadly uninhabitable) planet, you can orbit forever at an altitude of inches, passing over the surface at incredible speed.

You can cheat, a little bit, if you're not a really fast runner. Earth rotates at a pretty good clip – about 1,600 kilometres an hour. You can give yourself a boost by running with Earth's rotation, so run east. Most rockets do this, including the Space Shuttle.

Once you're in orbit, you're in strange territory, where faster is slower and vice versa. Say you want to pass over the surface of Earth more slowly. To do this you'll have to speed up. Try to increase your speed a little (how to do this is a mystery, since you're not in contact with the ground, and there's no atmosphere so flapping your arms won't help, but remember, we're not worrying too much about reality). As you increase your speed, your orbit will rise, but your speed over the ground will actually be a little slower (this is an illusion due to the rotation of Earth). Keep speeding up and eventually you'll rise to the altitude of the International Space Station. At that speed you'll be doing a complete orbit in just over an hour and a half – very slightly slower than you were going just above Earth's surface.

If you want to hover over Earth's surface, you'll have to go faster and higher still, boosting yourself out to about 35,800 kilometres. You'll be travelling in a circle around Earth at a very high speed. However, relative to the surface of Earth, you'll be locked in a fixed point in the sky. The situation is analogous to being on the outer edge of a vinyl record spinning on a turntable and looking at the label, which represents Earth. In space this is called a geostationary orbit, and it's a very useful thing. This parking orbit is where we keep pretty much all our communications satellites so that they always stay at a fixed point in the sky and we don't have to move our satellite dishes to follow them around.

So that's how to run into orbit. All in all, taking a rocket might be easier.

How much junk
is there in space?

Even the most distant reaches of our planet are haunted by the detritus we leave behind. Mount Everest is covered in discarded food wrappers, oxygen bottles, and human waste (as well as the odd human body). The most remote beaches in the South Pacific are littered with plastic water bottles, running shoes, rubber ducks, and other garbage that we've dumped, intentionally and accidentally, into the seas. It should come as no surprise, then, that those parts of space that we've visited are now filled with junk, trash, waste, offal, rubbish, and residue. While litter on Earth can be an environmental problem, it's often just an eyesore. Space junk, on the other hand, once discarded has a nasty habit of turning into potentially lethal hyper-velocity projectiles – perfect for smashing into spacecraft and posing a serious threat to astronauts.

Ever since we started launching things into orbit we've left junk in space. The oldest piece of space junk would be a museum piece if we could ever get it back to Earth. It's one of the earliest scientific probes, the *Vanguard 1* satellite, which was launched by the United States in 1958. It's been inoperative – and so basically junk – since 1964. It careens around Earth once every two hours just waiting for something to get in its way. There are also dead booster rocket stages up there, abandoned after they flew satellites into higher orbits. These boosters are flying bombs, as they often contain some unburned fuel that could spontaneously explode, turning one piece of space junk into many. Then there's the junk from human space missions, including lost tools, dropped nuts and bolts, and more disgusting stuff: astronauts on early Space Shuttle missions and cosmonauts on the Russian space stations used to routinely dump their garbage and even human waste overboard rather than return it to Earth. NASA put an end to that when the risks of doing so became clearer.

How much space trash is there? The U.S. Defense Department estimates that there is nearly 2 million kilograms of debris in low Earth orbit. On the one hand, that's not a huge amount – less than the mass of one fully loaded *Saturn 5* rocket, which is the vehicle that the United States used to get to the Moon. On the other hand, 2 million kilograms is also the same mass as 750 million .22 calibre bullets. Unfortunately, thinking about this stuff as bullets is all too appropriate, because most of the space junk up there is in small pieces going extremely fast.

NASA scientists track as much of this debris as they can with ground-based and satellite radar and telescopes. They have an eye on about ten thousand objects larger than 10 centimetres across (about the size of a grapefruit), including derelict satellites and old rocket stages. They keep a close eye on these because anything that size striking something like the Hubble Space Telescope or the International Space Station would probably destroy them. They can't track, but at least they try to survey, the roughly hundred thousand objects between 1 centimetre and 10 centimetres they think are careening around there. These are the bullet-sized objects that could easily damage or disable a space-craft or kill an astronaut. An uncountable number of scraps are smaller than that – things like paint flakes and bits of slag from solid rocket engines – which could still be dangerous but pretty much amount to orbital hail.

The reason this junk is such a problem is that everything in orbit is travelling at terrific speed. These speeds vary depending on the orbit, but speeds of 40,000 kilometres an hour aren't unusual, and that's something like ten times as fast as a rifle bullet. At that speed a very small object can be a very big threat to a fragile spacecraft or an even more fragile astronaut. According to NASA, a 1 centimetre diameter object travelling at that speed has the energy of a bowling ball hitting you at 500 kilometres an hour.

Collisions are far from unusual. The Space Shuttle, for example, is frequently hit. It has returned from space on more than one occasion with large chips in its windshield from a high-speed collision with a speck that might have been a paint chip or a bit of fuel slag. Had the shuttle window been penetrated by something larger it would have been disastrous. Most satellites are hit occa-sionally by small bits of debris, and larger impacts are thought to have disabled at least two orbiting satellites. The Hubble Space Telescope has experienced a number of hits, some of which have done significant damage to its solar panels. These collisions cause even more problems as they result in more fragments flying around. In January 2007, a test of a Chinese military anti-satellite system illustrated this problem. The Chinese fired a rocket inter-ceptor at an obsolete weather satellite and unfortunately the test was successful. A single, easily tracked derelict spacecraft was turned into a cloud of high-velocity space shrapnel.

Eventually most of this material will fall back to Earth. The orbit of space junk, like anything else in low Earth orbit, eventually decays, and the smaller pieces of junk mostly burn up on re-entry into the atmosphere. Sometimes this happens quickly. A spare glove lost on orbit by astronaut Ed White during a 1965 spacewalk probably re-entered a month or two later, as did a spatula lost from the Space Shuttle in 2006. An old Russian spacesuit tossed out of the International Space Station in January 2006 spent several months orbiting as a low-tech satellite before re-entering that September. For a lot of junk in higher orbits, however, falling to the level of the atmosphere can take a while. *Vanguard 1*, for example, isn't expected to re-enter Earth's atmosphere for more than 240 years. And with more space launches and more satellites going up every year, the problem of space junk is only going to get worse.

There's no real defence against these impacts, and the chances of a serious accident are pretty high. It's been calculated that during the life of the International Space Station there's a 1 in 91 chance that a spacewalking astronaut will be hit by a piece of space junk large enough to penetrate a spacesuit. The astronaut would be very lucky to survive.

WHAT DO YOU DO
WITH A USED SPACE STATION?

DESPITE THE EXPENSE OF BUILDING and launching things like satellites and space stations, we don't do much in the way of the three R's when it comes to disposing of them. We do reduce – space vehicles, because of the cost of their launch, are as light as possible, so they contain the minimum amount of material. Re-using and recycling, however, are for the moment not on. It's simply not practical to attempt to recover or re-use space stations and spacecraft, and so when we're done with them we throw them away. Throwing away a large spacecraft, however, can be a problem.

In the past we've simply abandoned them, leaving them like garbage in orbit. As a result, many of them are now hazards to space navigation and are tracked carefully to make sure they don't cross the path of a currently operating spacecraft and destroy it. This problem is compounded, but also partially solved, by the phenomenon of orbital decay. Discarded spacecraft don't just stay in their orbits. Over time, friction from the tenuous atmosphere that extends hundreds of kilometres into space will slow them down, and they will gradually drop toward Earth. Eventually they aren't going fast enough to stay in orbit, and they fall out of the sky. In most cases, that's the end of them. At about 80 kilometres up the pressure of their re-entry tears them apart, and then friction heats up the debris and vaporizes it. Occasionally, though, chunks of spacecraft survive the scorching heat of re-entry and fall to Earth. The surviving parts are, unsurprisingly, the heaviest and densest parts – which makes them the most dangerous bits – and they pose a risk to people on the ground. Satellites can also carry hazardous materials, in particular those powered by nuclear power. These are not generally used today in orbital missions, but have been in the past. When a spacecraft with one of these reactors re-enters there is a risk that the radioactive material in them will be released.

One of the best-known examples of this kind of mishap occurred in 1978 when a Russian spy satellite, COSMOS 954, re-entered and broke up over Canada's north. At the end of its useful life, the satellite was supposed to have been boosted up to a "disposal orbit," where it would stay safely in orbit for nearly a thousand years while its radioactive fuel decayed. The boost failed, however, and when the satellite re-entered, it scattered nearly 50 kilograms of uranium and other radioactive elements over a swath of northern Canada east of Great Slave Lake. Only a tiny amount of radioactive material was recovered in the cleanup, and much of the rest probably vaporized in the upper atmosphere, where it is now circulating. That wasn't the only nuclear accident in space. Both Russian and American nuclear spacecraft have had accidents that have resulted in the loss of their nuclear fuel. The radioactive material has vaporized in the atmosphere, been lost in remote mountains, or sunk in the depths of the ocean. None of these mishaps have resulted in contamination of any highly populated areas, and the amount of radiation released isn't thought to be enough to cause serious problems – so far no cases of radiation poisoning are associated with them. But the overall result is that our environment is a tiny bit more radioactive than it would be otherwise. We've been lucky. If one of these spacecraft accidents had resulted in nuclear waste being dumped on an inhabited area it would have been much, much worse.

A still serious but not quite so frightening concern is the potential risk when large spacecraft re-enter. Spacecraft are built light, but must carry robust parts, which can survive the heat of re-entry and strike the ground (or anything between it and the ground). When the Space Shuttle *Columbia* broke up on re-entry in 2003, for example, debris was scattered across much of the southwestern United States. Even human remains survived the re-entry. The Space Shuttle, of course, was designed to survive re-entry while most spacecraft are not. Even so, with large craft, big pieces can survive re-entry and pose a serious risk to people on the ground.

This has been an issue especially with the early space stations, which are among the largest things we've put in orbit. The first space stations were the Salyut missions launched by the Russians in the 1970s. The plan was to eventually "de-orbit" them in a controlled way by using rocket engines to precisely slow them in orbit. This would cause them to descend into the atmosphere and then fall to Earth. Ideally this de-orbiting would be timed to make sure most of the debris ended up in the ocean.

Several Salyut space stations were successfully dropped into the Pacific in this way, though two of them missed the ocean and deposited debris on the Andes. In 1991, *Salyut 7* rained down in small pieces of metallic debris on the town of Capitan Bermudez in Argentina, about 400 kilometres from Buenos Aires, but no one was struck or injured. In 2001, the Russian space station Mir ended its service life after fifteen years in space. Its de-orbit was most notable for how uneventful it was. Mir was slowed by a rocket burn and broke up over the Pacific, east of Fiji.

The U.S. space station Skylab had perhaps the most famous and nerve-racking re-entry. The Americans launched their Skylab station in 1973, following the Moon landings. The Skylab mission actually used a converted Saturn rocket stage for its lab space. This rocket stage had been designed as the booster that would send an Apollo mission to the Moon, but the structure was refitted to make a serviceable orbital space station.

NASA had studied several ways of controlling re-entry for this jury-rigged space station, but found none that was easily workable. In the end Skylab was launched with no real plan for its disposal. The station was abandoned nine months after it was launched, but was expected to stay in orbit for eight more years. This, NASA staff hoped, would allow enough time for the development and flight of the United States' new space vehicle, the Space Shuttle, which could then potentially boost Skylab up so it could be used again. In fact, the shuttle was delayed, and Skylab's orbit didn't last as long as was expected. By 1978 it was clear that Skylab was going to fall. This caused a low-level worldwide panic as it wasn't at all clear until the last moment where Skylab would hit. The station contained several large pieces of equipment, including a two-ton radiation-resistant vault used for storing film. Nobody wanted this dropping on their heads from orbit. Skylab might have caused a major disaster had it hit a city, and it was calculated that there was a 1-in-7 chance of its debris hitting an inhabited area. In the end, it reentered in 1979, scattering debris over relatively uninhabited western Australia. There were no human casualties, though there was an unconfirmed report that a cow was struck and killed. The town of Esperance in southwestern Australia keeps a display of debris from Skylab in its municipal museum to this day.

Having learned from Skylab, NASA does have plans for deorbiting the International Space Station, the largest object ever assembled in orbit, and thus the largest object ever dropped. When this will happen is, of course, currently unknown. The ISS is still being built, though it is years behind schedule, and at the moment its operating life depends more on budgets and politics than it does on orbital stability. In fact, the ISS falls toward Earth every day by about 30 metres due to friction with the wispy atmosphere at its roughly 400 kilometre orbit. To offset this, the ISS is frequently boosted up in its orbit by both the Russian Progress cargo delivery vehicles that deliver supplies to the station, and by the Space Shuttle when it docks with the ISS. The shuttle has boosted the ISS by as much as 40 kilometres on various missions.

The plans for taking down the ISS are elaborate, as it will be a complicated manoeuvre. Aiming something as large and ungainly as a space station takes some doing. In the best-case scenario, the debris from the station's break-up in the atmosphere will be dispersed widely over an area 300 kilometres wide and more than 5,000 kilometres long. Clearly a large empty space is needed. The intent is to drop the station into the Pacific, which is one of the largest, unpeopled spaces we have.

Now, having a plan is one thing. Being able to execute it is another, and there's some doubt that NASA currently has the ability to de-orbit the space station with the accuracy needed to keep the risk of injury to its goal of 1 in 10,000. A review committee of the prestigious National Academies of Science released a report in 2000 suggesting that NASA needed a more powerful and reliable propulsion module, with more fuel than is currently designed for the de-orbiting mission.

So until you hear more, keep an eye on the sky, and keep your head down.

We're in orbit.
Now what do we do?

CURRENTLY WE USE THE CHEMICAL ROCKETS to travel once we're in space that we use to get there, with the same disadvantages that rockets have for travel up from Earth's surface. Rockets require a lot of heavy fuel, and you have to haul that fuel around with you in order to use it for travel. That makes travel within our solar system – to Mars or the asteroid belt or the outer planets – slow and expensive. It makes sense, then, to look to other technologies that are more efficient and less expensive. We can choose from several new space propulsion technologies that have made it out of the pages of science fiction and into practical development, and even into testing in space.

The best developed technology, and one of the most promising ones for the future, is something that sounds like it should be in science fiction. It's called the ion drive. The ion drive has powered two robotic space missions – NASA's Deep Space One probe that visited a comet and the European Space Agency's SMART-1, which studied the Moon – and small thrusters built on similar principles have been used for years on Russian satellites to keep them properly in their orbits. The ion drive is a proven technology that could be a big part of the future of space travel.

The basic idea behind an ion drive is to do what chemical rockets do, only better. Chemical rockets are designed to shoot gas out of an engine as fast as possible. This propels the rocket in the other direction, simply because of Newton's basic law that for every action there is an equal and opposite reaction. To get a lot of thrust, chemical rockets burn a lot of fuel to make the gas very hot, so it shoots out of the engine as fast as possible. The faster the gas shoots out, the more thrust you get. There is, however, a limit. It's difficult to get a gas velocity from a chemical reaction beyond about 20,000 kilometres an hour. This is pretty fast but faster would be better.

Faster is what the ion drive does. The ion drive is a rocket that also shoots out gas for thrust, but it doesn't use chemical energy. Instead it uses electricity. The gas is ionized – stripped of an electron so it's given an electrical charge. Then the drive uses a powerful electrical field to accelerate the charged gas to up to 100,000 kilometres an hour and shoots it out of the back end of the drive. This results in five times as much thrust as you get in a chemical rocket. That's not the only benefit of the ion drive. Because the drive uses electricity, it can rely on many sources of power. Currently the best source, at least for smaller spacecraft, is solar energy. Solar panels harvest the energy of the Sun, turn it into electricity, and use that to drive the engine. You still need to carry gas to use for propellant, but you don't need a gas that's explosive. The gas is just what's called "reaction mass" (remember

Newton) and not an energy-rich fuel. Other sources of energy could be used as well. NASA is exploring the idea of using a nuclear reactor as a power source for an ion drive. There will, obviously, be important safety concerns with using nuclear power, and not just for the astronauts. Those nuclear reactors will have to be launched into space, and a launch accident would be a very bad thing.

The problem with ion drives, so far, is that they tend to be very small. Essentially, ion drives push out gas just a few atoms at a time, so while each atom gives a lot of thrust, the total thrust they provide is small. It's a bit like fuel-efficient cars on Earth – they don't have a lot of pick-up, but they can run for a long time on a single tank. In space this allows you to build up great speeds over time. They can achieve over weeks with slow acceleration at least the speeds that rockets give spacecraft in minutes with fast acceleration. They're very efficient, and they can get you where you're going, but they can't pass in the fast lane.

Like rockets, though, ion drives still have to carry propellant and engines. There are methods of travelling through space that involve no engines and no fuel at all, which is the ultimate in mass efficiency.

One option for accelerating things out of orbit is simply to throw them using a device that would resemble nothing so much as a giant magnetic sling. This technology is called an electro-dynamic tether. NASA flew a tether experiment on a Space Shuttle mission in 1996. The tether system siphons energy from Earth's magnetic field to generate electricity, which it uses for motion. The idea is quite simple, using a principle taught in high school physics labs – inducing an electrical current using a magnet. When a wire is dragged through Earth's magnetic field by a spacecraft, it's just like passing a wire over a magnet – an electrical current begins to flow in the wire. Harnessing the electricity for motion is the tricky bit, but in the end it uses almost exactly the same principle as is used in electrical motors to turn electricity and magnetism into rotation. The electricity can be used to spin the tether. If a weight is attached to each end, you end up with something like a rock being spun at the end of a string, with energy stored in the rotation.

To use this for launching something out of orbit, a spacecraft launched from Earth would dock at one end of a long spinning tether, balanced by a counterweight at the other end. The tether would then speed up its rotation, siphoning more energy from the Earth's magnetic field. The spacecraft would spin faster and faster at the end of the tether, and then it would be released. It would be flung away from the Earth into space like a stone shot from a sling. The great advantage of this system is that it harvests energy already present in the space environment and uses that energy to launch the payload.

Another free ride can be caught in space by using something called gravity assist, which is arguably the most successful space-flight technology we currently have. In a gravity-assisted flight a spacecraft already moving through space approaches a moon or planet and does a close fly-by. The gravitational pull of the planet accelerates the spacecraft as it approaches and, of course, slows the spacecraft as it travels past and away from the planet. However, if the spacecraft approaches the planet in the direction the planet itself is orbiting, it can get an extra boost in speed equal to the orbital speed of the planet itself. Several spacecraft, including the *Voyager* missions, which have now passed out of the Solar System, have used gravity assist very successfully to boost their speeds.

Like gravity assist, a solar sail also provides a way to get free energy, though in a rather different way. The idea behind a solar sail is to drag a spacecraft behind a very large, but extremely thin, reflective sail. The sail would catch photons of light from the Sun, and the pressure of these photons would push the sail. Photons are small and exert very little pressure. With a large enough sail, however, enough force could be captured to push a spacecraft around the Solar System. The Japanese have success-fully experimented with solar sail deployment in space. A private solar sail experiment called COSMOS-1, funded by the Planetary Society, unfortunately failed in 2004 when the Russian launch vehicle malfunctioned. A similar idea to the solar sail is the magnetic or plasma sail. The Sun doesn't emit just light but also charged particles that form the solar wind. A spacecraft gen-erating a sail-like magnetic field could catch the solar wind and ride it around the Solar System.

The most extreme space-propulsion technology might be a design that was abandoned in the sixties. It would allow for – in fact it requires – a large, heavy spacecraft capable of carrying dozens of people and huge loads of cargo. It would achieve ter-rific speeds, making trips to the planets matters of months instead of years. It's called nuclear pulse propulsion, and it seems a little insane as the idea is to ride the explosive power of a nuclear explosion. In a weird way this idea has a bit in common with the internal combustion engine found in every car. In these engines, a fuel-air explosion drives a piston, and the motion of the piston is captured (via a crankshaft) to turn the wheels. In the nuclear pulse engine there's an explosion – a half-kiloton yield atomic bomb – and a piston weighing a thousand tons. The piston is attached to giant shock absorbers, and on top of the shock absorbers lies the spacecraft. Propulsion is simple. A bomb is tossed behind the spacecraft. The bomb explodes, the piston is smashed by the tremendous energy of the explosion, and it is driven into its shock absorber, turning the violent explosion into a powerful push. The piston returns to its resting position, and another bomb is tossed out.

The designers of this system imagined a giant spacecraft with a crew of perhaps fifty and using thousands of bombs for propulsion. It would, in fact, take about six hundred bombs – one every half-second – to get the craft into orbit. From there it could putt-putt its way on a grand tour of the Solar System until it ran out of bombs. The advantage of this project is that there are no outstanding physics or engineering problems to solve in order to make it work. It could be built. It won't be, of course, and this illustrates the difference between the early enthusiasm of the nuclear age and our modern respect and fear of nuclear technology. The very idea of exploding six hundred nuclear weapons in the atmosphere, as would be required to launch an Orion spacecraft, is unthinkable to most people today. Who says the human race isn't evolving?

How do you cook
an astronaut?

Unless you have a taste in, well, let's call it "unusual" cuisine, you'll agree that what we really want to know here is "how do we *not* cook astronauts?" The dangerous radiation that blasts through space is one of the biggest problems we'll face if we pursue space travel beyond short jaunts into orbit.

Here's an illustration. In August 1972, between the *Apollo 16* and *17* missions to the Moon, a large solar storm occurred that spewed intense particle radiation into space. Had astronauts been in transit or on the Moon at the time, they would have received a huge dose of radiation. Even if it hadn't killed them during their trip, it would likely have made them terribly sick with radiation poisoning. They would have been vomiting and disoriented, with blistered skin and impaired vision and cognition, making it virtually impossible for them to pilot their spacecraft safely. Had they managed to return to Earth, they might have survived, with proper treatment, but their chance of long-term health problems like cancer would have been significantly increased.

While a disastrous event like this hasn't happened yet, there have been less dramatic experiences. During the *Apollo 11* mission to the Moon, and routinely during missions afterward, astronauts reported seeing flashes of light while their eyes were closed. This was because neurons in the retina or possibly even in the optic nerve itself were being struck by cosmic rays – particle radiation from the Sun and also from deep space. In October 1989 there was another solar flare that was even stronger than the potentially lethal one that occurred in 1972. It would certainly have killed any astronaut exposed on the Moon. As it was, in just a few hours it gave the astronauts on the Soviet Mir space station the same dose of radiation as a year's normal exposure in space.

A high-radiation environment in space is just part of reality and underlines how lucky we are here on Earth. The Universe is full of unshielded, uncontrolled, nuclear fusion reactors called "stars" that spray their radiation willy-nilly. We've got one of these dangerous objects right in the middle of our solar system – it's called the Sun. It emits a steady stream of radiation, including electromagnetic radiation from visible light to X-rays to gamma rays, but also sends out a constant stream of protons and atomic nuclei in the solar wind. On occasion it burps out larger doses when solar flares, storms, and phenomena known as coronal mass ejections occur. In addition to the radiation from our sun, there is also a steady stream of radiation from deep space called galactic cosmic rays. These are bursts of radiation from the supernova explosions of distant stars, which send streams of high-energy particles across the Universe at near-light speeds.

Earth's magnetic field and our atmosphere absorb a good part of the radiation from deep space and from the Sun. The result is that, for the most part, the only dangerous rays that make it to the planet's surface can be handled with a good SPF 45 sunscreen and a wide-brimmed hat.

As long as they stay in near-Earth orbit, astronauts are partially protected from radiation. Earth's magnetic field extends well beyond the orbit of the Space Shuttle or the International Space Station, lending some protection against solar and cosmic radiation. Both vehicles also have special radiation shielding, but even so astronauts routinely get something like twenty-five times the dose of radiation in space that they would experience from natural sources here on the ground. This is a significant health concern for astronauts, limiting the time they're allowed to spend in space and the number of missions they can fly.

Once we travel beyond orbit and the protection of the magnetosphere, radiation exposure is far higher, and the Sun's outbursts are that much more dangerous. Astronauts in the future are also likely to be exposed for much longer – a mission to Mars, for example, will involve a two- to three-year round trip. A colony on the Moon, which the United States has proposed establishing by 2020, would also have to deal with an unpredictable and dangerous radiation environment.

Shielding against radiation in space is difficult, mostly because to be effective it has to be thick and dense, which means heavy, and weight is the enemy of any space mission. Most space vehicles are made as light as possible, since every extra kilogram of mass launched into space increases the cost of a mission by at least tens of thousands of dollars. The *Apollo* Moon lander, for example, was constructed of such thin aluminum that a pencil could have been poked through its outer skin. The mission planners gambled that they'd be able to abort the mission and turn for home if any dangerous solar storm was brewing.

Since carrying lead plates into space for shielding isn't a good option, scientists are batting around alternative solutions for longer missions to the Moon and beyond. Colonists on the Moon might be best protected by a thick layer of lunar soil. This could be piled on top of buildings, or living and working structures could just be made underground. It's also been proposed that, with appropriate solar weather prediction, cocoon-like solar storm shelters might be practical, so that when the Sun was acting up, astronauts could hide like farmers in Kansas ducking into tornado shelters when a twister is coming.

Travellers on the long trip to Mars would have to bring their shielding with them. This could be in the form of just a thick-walled, heavier, and much more expensive spacecraft, but that's obviously not the preferred solution. NASA is now considering using water as a radiation shield. Water can be very effective in this – in fact, old nuclear fuel from nuclear reactors is stored in water, which keeps it cool and partially absorbs its radioactivity. Astronauts would require large amounts of water during the trip, so the idea, essentially, is to have them travel inside their canteen. By surrounding the crew quarters in a water balloon, it might be possible to have shielding while not bringing any more mass than you would have anyway. Another more speculative solution is to have the spacecraft make its own magnetosphere in the form of a smaller version of Earth's radiation shield. Using superconducting magnets powered by on-board generators, the ship could generate a magnetic field around itself that would deflect some of the most dangerous radiation. The ship would have to have significant electrical generating capacity to create the shield, but that could be supplied by a nuclear reactor – and nuclear reactors have been suggested as a propulsion system for a Mars mission. So the spacecraft's engine could double as its radiation shield generator.

Some solution will have to be found, otherwise radiation will be a show-stopper for future space exploration. After all, it's a fine way to prepare Buffalo wings, but no one wants their astronauts fried extra-crispy.

WHERE DID THE MOON
COME FROM?

PLAY THE HISTORY OF THE EARTH in reverse and you've got a Hollywood blockbuster. It has to be in reverse because the really big explosion comes at the beginning of the Earth's history, not at the end as it does in an action movie. There's another Hollywood similarity. The event that usually follows the explosion, in which the two romantic leads get together and form a couple, is also what happened with the Earth and the Moon. The Earth and the Moon have an odd romantic history. They're not like other planet-satellite relationships. They came together only because of a previous liaison between Earth and another body. That relationship was a total disaster.

That metaphor has been stretched quite far enough. Let's talk about just how the Moon was created, in one of the most incredible and catastrophic collisions in the history of our solar system. In fact, this is also the story of how Earth was created, because

Earth would not be what it is today without the event that created our moon.

The Earth-Moon system is an unusual one because the Moon is far too large to be a typical satellite. Moons in general – like those around the other planets in our solar system – are thought to have formed much as planets form. Back when the Solar System was just a large disk of gas surrounding a young star, planets formed by accretion – the dust clumping up into pebbles, then into rocks, asteroids, and finally forming a planet. Each young planet would have been surrounded by its own disk of dust as it formed, much of which would have fallen onto the planet, but some of which might have accreted in turn into moons in stable orbits in a miniature version of the same process that formed the planets around the Sun. This is probably how the many large moons of Saturn and Jupiter formed. Smaller moons may have formed elsewhere in the Solar System, drifting through space until captured by the gravity of a planet. This is probably how Mars acquired its tiny moons, Phobos and Deimos, which are less than 30 kilometres in diameter and are probably captured asteroids.

Neither of these mechanisms, however, could have been responsible for Earth's moon. The Moon is very large for a satellite – most moons are a tiny fraction of the size of the planet they orbit. The Moon couldn't have formed from a disk of material left over at the formation of the early Earth – it's too big. Models of how satellites form this way put strict limits on how big they can be, and the Moon is far bigger than that. The Moon also probably wasn't captured by the Earth. Among other reasons, it's just too big to be captured in any straightforward way by the Earth's gravity. The Moon is so large that it's sometimes considered not

Earth's satellite, but our binary companion – it doesn't orbit us as much as we orbit each other.

Until fairly recently, we didn't have a good explanation of just where the Moon came from. An explanation began to emerge early in the disco era of planetary science (the 1970s) when a radical new theory was suggested. The theory was that the Moon was formed from the debris of a huge collision early in the life of the Solar System between a rogue planet and the early Earth.

The early Solar System would have been a considerably busier place than what we see now. The planets were forming out of smaller chunks of material that had clumped together from the disk of dust and gas from which the Solar System had formed. The skies would have been filled with these kinds of chunks – asteroids and meteoroids and small planets, or planetesimals. In the four and a half billion years since then, most of this stuff has been gravitationally captured and swallowed up by bigger planets or swept away some other way, but this was early in the life of the Solar System before things had been tidied up.

The Earth itself was likely much smaller than it is today – possibly only two-thirds the size of the modern Earth. What's more, the Earth was not alone in its orbit. In an orbit not quite different enough from that of the Earth's, another proto-planet, roughly the size of Mars, had formed. Some astronomers have taken to calling this planet Theia, after the mother of the ancient Greek moon goddess, Selene. One fateful day these two planets came together in what must have been, even for this chaotic time in the Solar System, one of the most spectacular crashes that nobody would ever see. Theia hit the Earth from behind in its orbit, striking it at an angle. The impact melted both planets, but most of molten Theia bounced off at the impact only to be captured by the Earth's gravity and dragged back again, just as it was re-consolidating, again slamming into Earth with inconceivable violence in a kind of cosmic double whammy.

The end result of this was that the Earth, now entirely molten, absorbed much of the planet-sized mass of Theia, being born again as a new and larger world. It would have swallowed Theia's heavy iron core, merging it with its own, but some of the less-dense rocky material from the collision ended up in orbit around the Earth in a massive new ring around the planet. From this ring the new moon formed, possibly in as short a time as several decades. This new moon, still molten (as Earth still would have been), was probably fifteen times closer to the Earth than it is now, in an orbit just more than twice the diameter of the Earth, about 26,000 kilometres, and so it would have appeared far larger in the sky. Of course, had you been there you might not have noticed this detail as you'd have been distracted by the fact that the molten rock you were standing on was incinerating you. The tides would have been terrifically strong, raising bulges in the molten "ocean" of rock, and they would have happened much more frequently as well. The Earth's spin, accelerated by the impact, would have been far faster at that time, with a complete rotation – one day – taking just a few hours.

Earth started cooling quite quickly from this massive collision, and a solid crust would have started forming in just decades. However since the Solar System was still full of free-floating material, continued bombardment from space by comets, meteorites, and asteroids would have continued for a hundred million years or so, continually refreshing the molten surface.

As with many relationships, though, eventually, things between the Moon and Earth cooled and they've become more distant. This is not a romantic metaphor. The Earth and the Moon radiated away the heat from the energy of that great collision. The Earth's crust has cooled and solidified (thankfully), but it retains a very hot, largely molten core. The Moon, being smaller, has cooled almost completely and is now a solid ball of frozen rock with perhaps a tiny molten core. It has receded from the Earth over the last 4.5 billion years and currently orbits us 360,000 kilometres away. This was predicted by our understanding of orbital dynamics, but the gradual retreat of the Moon has been measured as well. The tides that the Moon raises on Earth essentially steal energy from the rotation of the Earth, slowing it down and lengthening our days a little every year (a billion years ago a day was eighteen hours long). This energy goes into speeding up the Moon, which causes it to orbit at a slightly greater distance. This slight change has been measured by scientists over the last nearly forty years by bouncing a laser range-finder off a small reflective plate the *Apollo 11* astronauts left on the Moon.

At the moment, the rate of that recession is a little less than 4 centimetres a year, but over billions of years that does add up. The Moon will continue to recede for many billions more years, but eventually it will stop as the Earth's rotation will slow to the point where it presents only one face to the Moon (as the Moon presents only one face to the Earth) and at this point the tides, too, will stop. This will end the transfer of energy from the Earth to the Moon, and so they will be locked facing each other until both are destroyed by the death of our sun. Like many couples, the Earth and the Moon may become a little more distant over time, but they'll stay together until the end.

How do we build
a Moon base?

President George W. Bush announced in September 2005 that he wants Americans to return to the Moon by 2020, and to establish a permanently inhabited Moon base by 2024. The Japanese space agency has also announced its intention to put a base on the Moon by 2030. Both China and India have publicized their ambitions to send some of their citizens to the Moon, at least for short visits. So if some or all of these ambitions are fulfilled, something like fifty years after Neil Armstrong's "small step," humans may walk on our satellite again, and perhaps this time we'll stay for good.

Given the extreme distance, expense, and difficulty of any trip to the Moon, we're going to want to make a lunar base as self-sufficient as possible. Bringing endless amounts of supplies and cargo from the Earth is simply going to be too expensive. For a Moon base to be sustainable, it's going to have to live as much as possible off the lunar land. This is going to be very difficult and will, even more than the challenges getting there, be what challenges our technology and creativity. ·

Let's compare, for example, the complexity of a Moon base to the most remote outpost we have on Earth, the Amundsen-Scott South Pole Station in Antarctica. The United States has maintained this station since 1956. The station can be reached only in the Antarctic summer, and then only by ski-plane. For six to seven months a year during the dark Antarctic winter, the station is cut off – it can't be reached by any vehicle or aircraft. Obviously the station has to be well stocked with fuel, food, and other supplies to get through all this, but in two essentials for survival it's self-sufficient: breathable air and water. We take these for granted on Earth, but on the Moon, even these will have to be supplied by rockets or somehow manufactured.

So how will we build a Moon base, and how will astronauts live off the land? They'll need a lot of things, but energy, shelter, air, water, and food top the list. One of the questions researchers are going to be investigating as they plan the development of a Moon base is whether any of these things can be extracted from the lunar environment.

Energy will likely be one of the simplest issues to resolve. There are really only two options. The first is nuclear power. We can make reactors compact enough to power nuclear submarines, so making a small one that could power a Moon base would not be out of the question. Nuclear power supplies such a consistent and large amount of energy that it will be a very attractive option, despite the risks of an accident in launching one from Earth.

The second and perhaps most obvious option for energy on the Moon is solar power, which we've used to generate energy in space from the earliest days of space exploration. Solar arrays on the Moon would be very efficient – far more efficient than they are on Earth because there are no clouds, dust, and atmosphere to reduce sunlight. However, the Moon rotates only once every month – a lunar day is about twenty-nine Earth days. A solar array just about anywhere on the Moon would be dark for a month at a time. Fortunately, there are two places where lunar solar arrays could produce power 100 per cent of the time – at the lunar poles. Because the Moon has only a small axial tilt, sunlight at the lunar poles is very nearly continuous. Earth's Arctic and Antarctic get the midnight sun for only a few months of every year, but the lunar poles see the Sun all the time.

Siting a Moon base at the lunar poles has other advantages. Because the poles are constantly in the Sun, they are the most temperate places on the Moon. Lunar temperatures at the equator, for example, vary between –180 and +100 degrees Celsius from lunar night to day. At the poles, the constant sunlight keeps things at a much more temperate and consistent –50 degrees Celsius. This is still cold, but not unmanageably so, and the more consistent temperature will be much easier on machinery and equipment.

There may be another important advantage to siting the base at one of the lunar poles: there may be water there. While the lunar poles have the most temperate climate, and the most sunlight of any place on the Moon, they also are home to the darkest and coldest places. The insides of impact craters at the poles, especially those with high steep walls, are in permanent shadow, because the Sun at the poles is at a very low angle in the sky. Because they're in permanent shadow, these craters are some of the coldest places on the Moon. Scientists think that this means they could harbour permanent water ice.

Water ice isn't all that uncommon in space. There's quite a lot of it locked up in comets, for example, and lots of small comets – and occasional large ones – end up hitting the Moon and Earth. They usually vaporize in the Earth's atmosphere, but on the Moon there's no atmosphere so they just smack into the surface. If the impact isn't too hard (which will vaporize the comet and cause the water to dissipate into space) ice can land on the lunar surface. On most of the Moon it quickly melts when the Sun hits it, but in a shadowed crater, it might be cold enough that it is preserved. This would be very useful to lunar explorers. The question is, is it really there? NASA's *Clementine* orbiter did radar imagery of the lunar south pole in the 1990s, which seemed to indicate that there was ice. This was such tempting evidence that in 1999, NASA decided to crash its *Lunar Prospector* satellite into the south pole to see if a plume of ice rising was visible after the impact. Unfortunately no plume was seen. At the moment, the evidence of water on the Moon is inconclusive.

The idea of finding ice is attractive not just because it would be a supply of water for astronauts to drink. It could also be used to grow food in a hydroponic garden. Water can also be cracked, using solar or nuclear energy, into hydrogen and oxygen. The oxygen would be useful for breathing, but perhaps more importantly, hydrogen and oxygen are valuable fuels. They could even be used to refuel lunar landers or to gas up rockets heading farther afield.

There is another resource on the Moon that could prove to be a rich source of all sorts of useful materials from air to fuel to building materials: the lunar soil. The top layer of the soil, or regolith, is a pulverized mixture of rocks and meteorites. For billions of years, carbon and metal-rich meteorites have been hitting the Moon, shattering themselves and smashing the rocks on the surface, creating a rich dust.

The regolith has all sorts of potential and could be useful in both crude and sophisticated ways. The crudest use is simply to dig into it to use it as a shelter. A lunar habitat buried under regolith on the Moon would be protected from small meteorite impacts and also somewhat insulated from cold lunar temperatures. A thick layer of Moon dirt would also provide some protection from solar and cosmic radiation, which will be one of the bigger concerns for humans on the Moon. It's also been suggested that structural materials can be made from regolith. Experiments have been done to try to melt the regolith with microwaves, to see if it can be fused into a brick that could be used in construction of buildings.

There are lots of useful minerals and chemicals in the regolith as well. It's rich in silicon dioxide, and iron and calcium oxides, potential sources of oxygen and usable metals. Iron is an extremely handy metal, of course, and silicon could be used to make relatively inexpensive solar panels. The regolith could be mined for material for the beginnings of lunar industry.

So the Moon, despite its sterility, might be able to provide many of the resources that a human settlement would need: water, air, energy, and shelter. The problem is that extracting them is going to require new techniques and technologies, since we've never done this kind of work in space before. It needs to be done efficiently and safely, and all this in one of the most hostile environment humans have ever explored. It makes Antarctica seem practically homey.

Can we turn the
Red Planet green?

As we look out into the Universe at other stars and solar systems for planets similar to Earth to visit, and perhaps one day to colonize, many scientists and space enthusiasts think we should stay close to home. Rather than attempting to travel the vast distances between the stars, a more practical proposition might be to transform relatively nearby Mars into a habitable planet. Now to be frank, to call something "more practical" in this context is to be, to put it mildly, disingenuous. It's not as if we have a choice between these two options. We currently have no technology to take us to other stars and even going to Mars is beyond us at this point. Given this, the idea that not only could we get there, but we could somehow master the technology necessary to transform Mars into a habitable planet – terraforming is the useful term for this – is somewhere between ambitious and preposterous. Nevertheless, scientists have been giving the idea some attention. All that distinguishes this idea from science fiction is the fact that there is literature on the subject in the form of scientific papers and not just novels.

What makes the idea even conceivable is that, as was pointed out in a previous answer to Question 2, Mars was once much more like Earth. We think that at one point in its early life Mars would have had an atmosphere much like Earth's early atmosphere, and it probably had another essential ingredient for habitability – water, and lots of it. Today Mars has only the most tenuous atmosphere – less than one-hundredth the density of Earth's – and while there are geological signs of water in the past, there's considerable debate about how much is there now. We can't see much, and what we can see is ice. But are the gases and water that gave the early Mars an atmosphere and oceans still stored there, and if so, can we release them?

Actually, the question of whether we could terraform Mars comes second to the question of whether we should terraform Mars. Mars may once have had life of its own, and for all we know it still has life of its own. Since in its early history it had conditions similar to those on the Earth, it's quite possible that simple organisms arose on Mars just as they did here. We know very little about how life started on our planet, and we mostly imagine it evolving in a peaceful little puddle bathed in temperate breezes and gentle sunlight. This may not have been the case. One plausible theory is that life started in extreme and apparently unlikely conditions. We've found complex communities of life around boiling-hot, but energy- and nutrient-rich, deep sea vents, as well as deep underground in warm rocks and in cold briny pools in the Antarctic. Apparently hostile environments like these may have been where life got its start. If life on Mars arose in these kinds of environments, it may still be there locked underground, possibly deep in the interior of the planet, not easily detectable by us from space. There was a tantalizing hint of such life when the *Mars Express* orbiter detected traces of methane in the thin Martian atmosphere in 2004. Methane can be a signature of microbial life. The problem is that if we were to attempt to transform Mars into an environment for humans, it could threaten these extreme but fragile ecosystems, or even extinguish existing Martian life. Many researchers in astrobiology, the discipline that considers life on other planets, think that if there's even a chance that such life still exists, it would be profoundly unethical to threaten it by terraforming Mars. Some scientists suggest that we might threaten the survival of Martian life just by visiting the planet.

An interesting complication here is that if there is life on Mars, it's possible that it didn't originate there. It might have come from Earth. Researchers have speculated that asteroid impacts on Earth could have sent large amounts of rocky debris hurtling toward Mars. That debris might have carried hardy micro-organisms all the way to the Red Planet, and they might have colonized it. Thus we wouldn't be endangering alien life forms, but our own microbial offspring. It's also possible that this process worked the other way. Life on Earth could have come from Mars. In that case terraforming could threaten the existence of the descendants of our original ancestors.

If we brush those considerations aside, we're still left with the question of whether we actually could terraform Mars. The answer depends initially on whether the raw materials are still there. It would be too difficult to bring an atmosphere or any significant amount of water to Mars, so it's essential that these things are still there – at least partly hidden away. We have seen some of the material that could form the atmosphere. Orbiting probes have confirmed that the polar ice caps on Mars contain large amounts of frozen carbon dioxide, and there's probably much more CO_2 frozen into the soil. If the temperature could be increased on Mars enough to melt this CO_2, it would be an excellent start on an atmosphere. There have been various proposals for how to do this. One suggestion is to build chemical plants on Mars to produce super-greenhouse gases like fluorocarbon compounds, which are extremely efficient at trapping heat. This would cause Mars to retain more of the Sun's energy, gradually heating the planet. The frozen CO_2 would then begin to melt and begin to re-establish the atmosphere. Of course CO_2 is a greenhouse gas itself, and so as it melted it would help to warm the planet, accelerating the process in what scientists call a feedback loop. Huge solar reflectors in space around the planet, redirecting sunlight onto the polar ice caps, could accomplish the same goal. Three such reflectors, each the size of Alberta, would do the trick.

The next step would be to try to restore Mars's oceans. Once the CO_2 melts, and the greenhouse effect begins to warm the planet in earnest, Mars should begin to release its water. Not that we know how much water there is on Mars. We know there's some, perhaps quite a lot. In 2007, results from radar surveys revealed that the south polar ice cap is up to 4 kilometres thick. If the greenhouse effect could warm the planet enough to melt the ice cap, and possibly liberate the water we think is frozen into the Martian soil, the process of terraforming would be well on its way.

The atmosphere would need more work, though, and nitrogen might be a problem. Most of our atmosphere on Earth, about 70 per cent of it, is composed of nitrogen. It's essential to life, but Mars's atmosphere is only 3 per cent nitrogen. If Mars has no nitrogen in the minerals of its crust, then the only way to terraform would be to bring it in. This might be possible – there could be nitrogen-rich asteroids in the asteroid belt beyond Mars that could be "driven" with huge rockets to Mars and dropped on the planet. This, however, would turn an already nearly inconceivably complex project into something even more ambitious.

In the best-case scenario, if water and nitrogen are still present in large quantities, the process of heating and releasing it into the ecosystem could be quick, and not just on geological time scales. It's been suggested that it could take as little as a hundred years. At this point, the planet would be something of a blank slate, much as Earth was before life arose. It would still be a hostile environment with an atmosphere of mostly nitrogen, CO_2, and water vapour. Humans might be able to survive there without pressurized spacesuits, but we'd still need an oxygen supply. The oxygen in Earth's atmosphere was largely produced by photosynthetic life forms, so seeding Mars with simple microbial life, like blue-green algae, would be the next step. These microbes would take sunlight and CO_2 and produce oxygen. As these organisms built a more oxygen-rich environment, more complex plants could be introduced to accelerate the process.

One big problem with growing plants on Mars is one we don't face on Earth – radiation. The Earth benefits from multiple layers of radiation shielding. Our ozone layer in the stratosphere filters out much of the Sun's ultraviolet radiation. We're also surrounded by a powerful magnetic field that protects us from the heavier radiation of the Sun called the "solar wind." This is a stream of high-velocity charged particles, which would be lethal to us on Earth if our magnetic field didn't deflect much of it. Mars has no ozone layer and only a feeble magnetic field, which would offer little protection against hard solar radiation. Radiation would damage all unprotected life on Mars and it could prevent life from taking hold on the surface. It's not clear what the solution to this problem is, though breeding or engineering radiation-resistant micro-organisms and plants is one part of it.

If the plants thrived, it could be as little as a thousand years before oxygen levels in the Martian atmosphere grew to something close to what they are in the high mountains on Earth, where humans just barely survive. The next job would be finding human volunteers to colonize the planet. It might be a difficult task, though, to find people willing to move to Mars. After all, if we had the wisdom and resources to turn the Red Planet green, we would also, presumably, be capable of the much easier task of cleaning up our act on Earth.

WHAT WILL HAPPEN WHEN
AN ASTEROID HITS?

THE CHANCE THAT YOU'LL DIE as the result of an asteroid striking Earth isn't huge, but it's higher than you might think. Scientists have put the number at around 1 in 20,000, about the same odds of dying in a passenger jet crash, and maybe a little higher than the chances of dying in a mobile home when a tornado hits. In fact, your chance of dying in an asteroid impact is far greater than the odds that the lottery ticket in your pocket will win you the multi-million-dollar jackpot this week. So, if the asteroid/plane crash/tornado makes it unlikely you'll be around to enjoy your winnings, why bother buying tickets?

When comparing asteroid impacts to plane crashes or tornadoes, however, the stats aren't helpful. They're comparing apples and oranges, or rather tragedies and catastrophes. In a tornado or a plane crash, anywhere between a few and a few hundred people might lose their lives. In a large asteroid impact, the numbers could be in the millions, possibly the billions. That's the real statistic to be worried about. There's a small chance that a large asteroid will hit the Earth in your lifetime, but if it does, you will most likely be in a very large line-up at the gateway to paradise.

Just how long that line-up would be depends on many variables. Like a tornado hit, an asteroid strike could be a relatively minor incident. However, there is a frightening possibility that it could also result in a global catastrophe that could end our species and many others. It depends entirely on what kind of object it is that hits us, how fast it's going, how large it is, and where it hits. All this will determine not just the number of lives lost, but just how people are going to be killed. No two asteroid impacts are going to be exactly alike.

Let's take one scenario that a group of astronomers from the University of California have analyzed. They looked at all the asteroids currently being tracked in their orbits and picked the one with the biggest chance of hitting us, 1950DA. It was first discovered in 1950, but then astronomers lost track of it. They found it again on New Year's Eve, 2000. It's about 1.2 kilometres across, and while we're not sure just what it's made of, chances are it's mostly rock, as most asteroids are. Its orbit is a large ellipse that goes from outside Mars's orbit to inside the Earth's orbit once every two-and-a-bit years. Since it crosses our orbit twice in doing so (once going inside our orbit, once going back outside), there is a chance it could strike the Earth. Astronomers have calculated that on March 16, 2880, there's about a 1 in 300 chance that the Earth and 1950DA will attempt to occupy the same space at the same time. It's hard to be more precise because it's a relatively small object and its orbit will likely be perturbed a little between now and then in ways that are hard to anticipate. For the sake of our exercise, let's assume the worst, that we're on the bull's-eye.

Just in case you're planning, by the way, it's going to be a Saturday.

1950DA is going to hit the atmosphere at a little over 61,000 kilometres an hour and will strike the Earth seconds later. We don't know exactly where it will strike, but the astronomers who ran the scenario decided upon the Atlantic Ocean about 500 kilometres off the coast of New England. The asteroid will dive through 5 kilometres of ocean and hit the sea bottom. The energy of that impact will be the equivalent of 60 billion tons of TNT, or about a thousand times more power than the most powerful nuclear weapon we've ever constructed. The impact will create a temporary hole in the ocean nearly 20 kilometres across. Unless there's someone unlucky enough to be in a boat in the vicinity, no one will witness the impact or notice any immediate effect. However, roughly a minute and a half later people living on the eastern coast of North America will feel a mild rumbling like a distant earthquake. About twenty-five minutes after the impact a shock wave of air will hit the coast, in the form of a sudden wall of wind and a loud rumble. Windows may be broken. In less than two hours, larger than normal waves begin to hit the shore. This will begin a series of tsunamis that will devastate the coastline. The waves will keep growing larger and larger, hitting every couple of minutes, until the peak wave, a monster 120 metres tall, hits land. The coastline will be scoured clean. Every North American coastal town or city will be completely destroyed, and the water will wash inland dozens of kilometres. North Americans will not be the only casualties. Eight hours after impact, 10 to 20 metre waves will pound the west coast of Europe and Africa, destroying most coastal settlements there as well, though with slightly less shocking violence. The death toll will certainly be in the many millions, and the economic cost will be incalculable.

Now here's the thing. This is a smallish asteroid. While millions would be killed, this isn't a big enough impact to lead to a global environmental catastrophe that might threaten the human race. It would be economically devastating, but it probably wouldn't threaten civilization. For that you'd need something bigger.

So let's look at another scenario. Let's say a 10 kilometre diameter asteroid were to strike in the middle of Quebec, about 500 kilometres northeast of Montreal. This location isn't just a random pick – an asteroid has already hit here. About 200 million years ago a 10 kilometre asteroid created a 100 kilometre diameter hole in what would be named Quebec, and the sign of this collision can be easily seen in satellite images of the strange circular shape of Lake Manicouagan.

Imagine yourself strolling down Rue Ste.-Catherine in Montreal enjoying the sights. The asteroid strikes, driving nearly 21 kilometres deep into the rock of Canadian Shield with the energy of 10 million mega-tonnes of TNT. Nearly 1,500 cubic kilometres of rock is melted or vaporized, and roughly half of it is ejected out of the crater into the sky, some of it even into space. A massive fireball appears on the horizon, nearly fifty times bigger than the Sun, and just seconds after the impact everything around you explodes into flame. Your clothing ignites, the flash of heat gives you third-degree burns over most of your body. Trees and grass burst into flame. A minute and a half later, an earthquake rumbles through, destroying the most fragile buildings but leaving most of the city standing, if in flames. A few minutes later, the sky starts to fall. Hot chunks of rock varying in size from grains of sand to boulders begin to rain down, burying the city's streets to a depth of a full metre. Less than half an hour after the impact, the air blast from the impact arrives – a super-hurricane with winds of 1,600 kilometres per hour, which knocks down the buildings still standing. It's possible that some people would survive in underground refuges, but there would be no one to rescue them.

Farther away from Montreal, the damage would still be intense. In Toronto, the heat flash from the fireball is nearly as destructive, as buildings and people burst into flame. The earthquake and the air blast will be slightly less destructive, though many buildings will still collapse. The rain of ash and debris will be smaller, only about 13 centimetres deep, but it will still be hot enough to start or maintain fires. Most of the population won't survive. Beyond the cities, the rain of hot debris will likely start wildfires across the continent, though the far west will not suffer so greatly from the immediate impact. Dust in the atmosphere will spread worldwide, though, blocking the Sun and causing crop failures and mass starvation. The long-term climatic impact will not be great, so the possibility of recovery is there. Human civilization, though, will be devastated, and it's possible that our technological society may collapse, returning us to a more primitive and desperate existence.

This isn't even the worst-case scenario. In a way, a catastrophe like that would be a lucky break. The Manicouagan impact, though devastating, didn't spawn a global extinction as far as we know. That's because it hit in hard, crystalline rock, away from the ocean. There was no tsunami. The rock that was shattered and vaporized didn't contain sulphur or nitrogen or carbon compounds, which, if vaporized, could lead to a generation of acid rain and millennia of climate disturbance. The wildfires created by the impact weren't global. Of course, change all that and what you have is the kind of impact event that wiped out the dinosaurs. If that happened today, it might well lead to the extinction of our species and many others.

Fortunately those kind of hits are rare. Impacts by 10 kilometre wide objects, which could cause global catastrophe and mass extinction, happen only every 100 million years or so. A collision by something around 1 kilometre across like Asteroid 1950DA can be expected more often, about every 70,000 years or so. Also, the ocean strike and the massive tsunamis are the worst-case scenario there. If a 1 kilometre sized object hit in northern Quebec it would be at worst a regional disaster. The hard rock the asteroid would strike would limit the damage, and while it would destroy everything within, say, 300 kilometres and set a good chunk of the province on fire, the citizens of Montreal would be spared. In fact, the luckiest place for an impact might be somewhere like the middle of the Sahara, where we'd barely notice it at all.

Given all this, maybe it's worth buying that lottery ticket after all.

How do you loosen

the asteroid belt?

Don't look now, but we're missing a planet. All we've got left is the pieces, and we're losing those.

A map of our solar system that gives the relative distances of the planets will show you what we're talking about. You'll notice that there is a pattern (humans are wonderful at seeing patterns): the distances from the Sun to the planets grow in a regular way. If you look at the orbit of any particular planet, the distance to the orbit of the next planet out is about twice the distance from the orbit of the next planet in (there's a little fudge-factor). This mathematical pattern works as far out as Uranus (but fails for Neptune and Pluto), and it was recognized as far back as 1766. It's known as "Bode's law," after one of the first astronomers to recognize it. Most astronomers today don't think it's terribly important, since all it really means is that you can't pack planets in too close or their gravity would cause them to crash into each other.

Nevertheless, it does reveal an interesting puzzle. Bode's law predicts the location of the planets – except there's one missing. When applied to the inner parts of our solar system, Bode's law seemed to indicate that there should be a planet between Mars and Jupiter, somewhere around 420 million kilometres from the Sun. Instead there's a gap – well, not so much a gap as a bunch of gravel. Instead of a planet, we've got the asteroid belt. When the first asteroid in the belt was discovered in 1801 by an Italian priest, it was initially thought to be the fifth planet. It was given the name Ceres, but not long afterward other objects in similar orbits were discovered. It became clear that there were in fact many objects in the sky similar to Ceres – a whole family of small planetoids at roughly the same distance from the Sun – and so the asteroid belt was discovered.

We know now the asteroid belt consists of hundreds of thousands of objects, varying in size from rocks to dwarf planets. Ceres is the largest of these at nearly 1,000 kilometres in diameter, a little less than a third the diameter of our moon, and twenty-six others larger than 200 kilometres in diameter have been discovered. The vast majority of the asteroids, however, are much smaller, though it's expected that there are probably as many as a million that are at least 1 kilometre across. Our common picture of the asteroid belt (largely gathered from science fiction movies) is of a dense, chaotic band of whizzing rocks. In fact, the asteroid belt isn't very dense at all. The total mass of all the asteroids in the belt is thought to be less than the mass of our moon, and while this makes a lot of small pieces, they're spread out over the vast area of space between Mars and Jupiter. If you were to stand on the surface of one of the larger asteroids, it's unlikely that you could even see any others with the naked eye. NASA has sent several spacecraft right through the asteroid belt without any real danger of collision, and missions to the asteroids, like the NEAR mission, which orbited and then landed on the 30 kilometre diameter asteroid Eros in 2001, have a tricky navigation job even finding their target.

So why is it that we have an asteroid belt – a pile of rocks – instead of a planet? Well, at one time it was thought that indeed there might have been a planet that at some point in the Solar System's history was smashed in a huge collision, perhaps with another rogue planet or giant comet or some such thing. These days, however, this idea has been more or less entirely abandoned. The distribution of the asteroids in the belt doesn't support the notion of a planetary explosion. It's also clear that most of the objects in the belt are primitive "chondritic" meteorites. These meteorites have formed from the original dust of the Solar System and have not been melted or crushed as they would have been if they'd once formed part of a much larger body. What we see in the asteroid belt today is material from the early Solar System that never formed a planet. At one time there was probably much more of it and the belt would have been much thicker, but much of its material has been lost over the several billion years of the Solar System's life. The asteroid belt is growing thinner all the time.

Jupiter seems to be to blame for this. Since the beginning of our solar system, this big bully has been pushing around the smaller kids in this part of space. As the Solar System was forming, this region of space would have had a great deal of material – anywhere between ten and one thousand times the mass of the Earth. As with other planet-forming "rings" around a sun, the material would have started as a band of gas and dust that slowly began to clump together, accreting into larger objects. In the rest of the Solar System, this process continued as bigger and bigger objects formed in each orbital band and joined together, finally forming planetoids and then single planets. In the region of the asteroid belt, however, the young Jupiter's gravity kept interfering in this process. Computer models of planetary formation suggest that the large mass of Jupiter (even as it was forming itself) kept disturbing the orbits of the planetoids in the asteroid belt. As a result, they never settled into the peaceful orbits that would allow them to continue to come together. As Jupiter's huge mass drifted by, they would be deflected, running into each other and on occasion, smashing to bits, but at other times being slung either toward the Sun or away from it, possibly right out of the Solar System. Gradually, much of the material of the asteroid belt was lost, either to deep space or to the Sun and, in rare cases, in collisions with other planets. Eventually the asteroid belt became the thin, relatively empty band of debris it is today.

We should be grateful for this in one way, and frightened of it in another. It's entirely possible that life on Earth was a result of this process. When Earth formed it's thought that the tremendous heat of the early molten planet would have driven off into space most of the water that would have been present during its formation as a planet. That water had to be replenished somehow for life to thrive. Initially, astronomers thought that much of the water we have on the planet today came from comets from the distant reaches of the Solar System colliding with Earth. Chemical evidence is now suggesting that in fact it came from the asteroid belt. Asteroids, whose mass was perhaps up to 10 per cent water, flung at us by Jupiter's gravity may be the major source. We may be benefiting from an unborn planet's stolen oceans. Asteroids in the belt are still being disturbed by Jupiter throwing its weight around, and it continues to throw the occasional snowy rock at us. One of these days that disturbance might result in Earth being hit by something large, hard, and extremely destructive. We will have some warning, though. Many of the asteroids we've identified that come close to Earth were shoved into their orbits by Jupiter, and it takes at least tens of thousands of years for even Jupiter's gravity to push an asteroid toward us.

If we discover that Jupiter has shoved something large at us, then some of our recent robotic missions to asteroids will prove their usefulness by allowing us to generate a plan to redirect an incoming one. We've learned, for example, that many asteroids are not solid rocks, but more like compacted rubble-balls that are likely quite fragile. This suggests that plans for deflecting an asteroid with something like a nuclear explosion probably won't work. The soft asteroid would absorb the explosion like a pillow absorbing a punch. Instead, we will have to find other, more gentle ways to persuade asteroids not to hit us.

Jupiter is continuing, slowly, to deplete the asteroid belt in other ways as well. Evidence of this appeared in 2004, when scientists found the remnants of a collision that happened a mere half-million years ago between two asteroids that were likely disturbed from their regular orbits in the belt by Jupiter and smashed into each other. The larger of the two was probably 15 kilometres across, and astronomers speculate that as much as half of its mass was ejected from the asteroid belt into space. Some of it doubtless ended up hitting Earth.

As an aside, we now know that our asteroid belt is not unique in the Universe. There was considerable excitement in 2005 when a much thicker belt was found in another solar system entirely. Astronomers detected a cloud of warm dust around a star called HD69830 roughly 41 light-years away, which they interpreted as the dust from numerous asteroid collisions in a denser asteroid belt than ours. Interestingly, the belt around this sun is about at the orbit of Venus, suggesting that a giant planet considerably closer to that star than Jupiter is to ours is doing most of the damage in that solar system. This rules out the possibility of finding an Earth-like planet in this system. Any Earth-like planet would be right between the bully and its victims, and while we can assume an Earth-like planet would be a plucky fighter, it wouldn't last long going up against a gas giant bully.

Who left those rings around the planet?

Planetary rings are among the most spectacular and beautiful things we can see in space. The rings we're most familiar with, of course, are Saturn's – the huge, shining, varicoloured concentric circles around the Solar System's second-largest planet. They're one of the sky's most wonderful sights. See them while you can, though, because they're fading and soon, in astronomical terms, they'll be gone.

Saturn's rings were first spotted in 1610 by Galileo, using his new telescope, and even today it's one of the great thrills of the backyard astronomer when Saturn finally is brought into focus, and the marvellous rings appear. Astronomers have been studying them now for four hundred years, but they still hide many mysteries. We are still discovering new things about Saturn's rings, but what's more surprising is what we're finding out about other planetary rings. One of the astonishing findings from the exploration of the outer planets is that all the large planets of our solar system have rings. This is remarkable because those rings shouldn't be there.

When Galileo first saw the rings of Saturn he didn't recognize them for what they were. His first telescopes weren't powerful enough to see the rings clearly. What he saw were bumps on either side of Saturn that he thought might be huge moons, almost half the size of Saturn itself. When he observed Saturn two years later, he was stunned that these objects had disappeared. What had happened was that the relative position of Earth and Saturn had put the rings edge-on to the Earth – and so they were invisible. Roughly forty years later, Christiaan Huygens realized that what Galileo had seen was actually a large flat disk around the planet, which turns edge-on to Earth roughly every fourteen years. It was a remarkable insight from someone using a telescope considerably less powerful than can be bought today in any department store. Only fifteen years after Huygens's discovery, the astronomer Cassini was the first to discern that the disk contained separate rings – he spotted two – and a third ring was discovered in the nineteenth century. It wasn't until then that James Maxwell deduced that the rings weren't solid disks but were made up of small particles – a finding that was confirmed first by powerful ground-based telescopes and then, more than a century later, by spacecraft.

From the data and pictures collected by spacecraft visiting Saturn, we've recently learned a lot about the rings. They're remarkable not just for their beauty, but even more so for their size. Saturn itself is about 120,000 kilometres in diameter. The innermost ring begins about 6,000 kilometres above the planet and the rings extend to about 480,000 kilometres from the centre of the planet. This is a huge distance. If these rings orbited Earth, they would continue out well past the orbit of the Moon. They are wide, but they are shallow, on average only about 1 kilometre thick. Given their breadth, this is incredibly thin – a sheet of paper scaled up to the size of the rings would be about three hundred times thicker. This means that the overall mass of the rings is relatively small – probably the equivalent of the amount of material in a moon not more than 200 kilometres in diameter. The rings seem to be made up mostly of small particles of ice and rock, ranging in size from a grain of sand to small ice boulders perhaps 10 metres across.

The first spacecraft to visit Saturn and get close-up images of the rings was *Pioneer* in 1979, followed quickly by *Voyager 1* and 2 in 1980 and 1981. That's when the first surprise in a hundred years about the rings was revealed. Saturn turned out to have not three, but seven, distinct, concentric rings. They're separated by bands of relatively empty space, which seem to be kept clear of material by the gravitational effects exerted by some of Saturn's many moons and moonlets. These have come to be called the "shepherd moons" for the way they guide and constrain the rings as they rotate around the planet. Each ring orbits at a different speed, so they move relative to each other in their rotation. From Earth, the subtle structures of the rings aren't easily visible, but when *Voyager* started sending back images in the 1980s, we could see remarkable detail in the features of the rings. Each of the seven large rings turned out to be formed from multiple "ringlets," which were really rings in themselves. From Earth the rings look relatively uniform in colour, but NASA's highlighted pictures from *Voyager*, which took images in infrared and ultraviolet light, showed tremendous variations in colour and some dizzying patterns. Admittedly, many of these "colours" are synthetic – false colours chosen to represent frequencies of light the human eye can't see – but they help reveal the delicate and detailed structure of the rings. We received yet more beautiful images of the rings when NASA's *Cassini* probe arrived at Saturn in 2004, and it continues to send back remarkable pictures.

We are extremely lucky to be looking at Saturn's rings, because it's very likely that, had humans appeared on Earth 150 million years ago or had we waited 150 million more years before looking at the sky, we would see Saturn as only another distant gas giant like Jupiter. The characteristic rings would be nowhere to be seen. The rings, we're coming to understand, are a temporary decoration for Saturn. As soon as they formed they were already on their way to extinction.

Very early in its life, a planet may have a ring made up of the material left in the disk from which the planet formed. This material, if it's close to the planet, will soon either drop down onto the planet or, if it is farther out, coalesce to form a more substantial satellite – a moon. Astronomers think rings can also result from catastrophes. The Earth probably had a substantial ring for a while after it was struck by a proto-planet 4.5 billion years ago. That material, though, either fell to Earth or was gathered in by our coalescing moon. So the natural situation of an undisturbed planet is to be ringless.

About 100 million years ago, this was likely true for Saturn. It was ringless, but may have had one or a few small icy moons in close orbits – inside what's called the Roche limit. These moons would have led a tenuous existence, as inside the Roche limit is a dangerous place to be. It's defined as the region near a planet within which the tidal forces exerted by the gravity of the planet are strong enough to tear a moon or asteroid apart. In fact, moons can't form inside the Roche limit, and Saturn's moonlets probably started out as either asteroids captured by the planet as they drifted through space or normal moons pushed inward by gravitational interaction with some of Saturn's multitude of more distant moons (somewhere between thirty and sixty at last count). They may have survived for a while, but then a passing asteroid or comet hit one of these icy bodies, or possibly they collided with each other, causing their disintegration, which was accelerated by Saturn's tides. Following this smash-up, over at least thousands of years, the debris settled into the flat structures of the rings, pushed into place and order by the gravity of the shepherd moons, and Saturn's magnificent rings were born.

The future of the rings, however, is not bright – in several senses. It's certain that the spectacular beauty of the rings will dim over time. One of the reasons we know the rings are relatively new is that the ice they are made from is still white and reflective. Over time, micrometeorites from interplanetary space will smash the ice crystals and what remains will be fragments of black rock and metal and a dirty dust of ice. The rings will become dingy and grey. This process will also gradually eat away at the rings, as these collisions will also knock particles of material out of their carefully constrained orbits, allowing them to be captured and dragged down to Saturn. The planet and its moons are fighting back, to some extent. Saturn's moon Enceladus has active volcanoes that are spewing ice out into space, and some of this is replenishing one of the rings. Over time, though, inevitably, the rings will grow darker and thinner and in perhaps a few hundred million years be nothing but a faint grey wisp.

These feeble grey bands are, in fact, the kinds of rings the other giant planets have. Jupiter, we now know, has three rings. This was a great surprise to astronomers when *Voyager 1* sent back pictures for the first time. The rings aren't as substantial as

Saturn's. In 1973 and 1974, the *Pioneer* spacecraft actually flew right through them without the mission controllers even noticing. Jupiter's rings are almost as wide as Saturn's, but they're dark and far less substantial. They didn't have a dramatic origin like Saturn's, either. They likely didn't form from any collision but rather from dust sprayed upward into orbit by the constant stream of meteorites that hit the giant planet's inner moons. These rings don't get thicker because the dust of which they are made eventually falls down onto Jupiter, so they erode as fast as they're built. Neptune and Uranus also have faint, thin rings, which are similarly nowhere near as spectacular as Saturn's.

When Saturn's rings finally do disappear it will, sadly, cease to be the jewel of the Solar System. That mantle could be taken up in time by two of the more distant planets. Both Neptune and Uranus have the raw material for new rings in the form of a close and fragile moon that could break up. Given a big enough collision, a new ringed planet could emerge. Let's leave it as a potential project for those interested in sprucing up the Solar System in the future.

WHAT MOONS ARE WORTH VISITING?

No GUIDEBOOK IS COMPLETE without a list of the most interesting places to visit while you're there. A guidebook to our solar system would have to include a section on the moons, which are in many ways far more interesting than most of the planets. In particular, the moons of the giant planets – Saturn and Jupiter – are more attractive destinations than the planets themselves. Both planets have such tremendous atmospheric pressure and massive gravity that they'd likely appeal only to tourists with suicidal tastes. Their moons, however, are fascinating places. While not actually clement, they might be possible to tour, given a good spacesuit and a stout pair of boots.

Let's take our planets in order, from the Sun outward, and survey those moons worth considering for your itinerary. We can quickly forget about the first three planets, Mercury, Venus, and Earth. Mercury and Venus have no moons, and we've already visited Earth's. Why Mercury and Venus have no moons is not entirely known. It might be that small inner planets like Venus, Earth, and Mars don't normally have moons. Earth's moon is clearly an oddball – the result of a huge collision early in the history of the Solar System. So perhaps Venus's moonless condition is the norm, not the exception for a planet that's close to its star.

This brings us to Mars, which has just two moons, Phobos and Deimos, Greek for "fear" and "panic." In Greek mythology, Phobos and Deimos were the sons and attendants-in-battle of Ares, the god of war, whom the Romans later named Mars. Both moons were discovered in 1877. As moons go, they're rather unimpressive, at least in size, and neither of them is spherical. Deimos is the smaller of the two, and it's only about 15 kilometres across at its widest dimension. It orbits about 23,500 kilometres from Mars. Phobos is both larger and closer, about 27 kilometres at its widest, orbiting about 9,400 kilometres from Mars. Phobos and Deimos are quite interesting in other ways than their appearance. It's not entirely clear where they came from. They appear to be similar to asteroids, and so the leading theory for their origin is that they were pushed out of the asteroid belt by Jupiter's gravity and were later captured by Mars. It's also been suggested that they could be the remains of a huge impact on Mars that ejected mass into orbit, which later accreted to form the moons.

Visiting Phobos and Deimos would offer all sorts of entertainment. Both are so small that their gravity is relatively tiny – less than a tenth of a per cent of Earth's. It would be possible to jump into orbit on Phobos or to throw a ball (if you had a good arm) that would completely escape the gravitational influence of the moon. If you did manage to keep yourself on the surface (take baby steps), you could see Phobos's sights. It's heavily cratered from meteorite impacts, and its most spectacular feature is the biggest crater of all, a cavity 10 kilometres across (remember, Phobos itself is only 27 kilometres across). The hole, called Stickney crater (after the wife of the discoverer of the two moons), looks as if someone took a huge ice-cream scoop to one end of Phobos and is likely the result of Phobos being smacked by an asteroid that must have come close to shattering the moon. There is some evidence, in pictures from the Russian *Phobos 2* satellite, that there might be water frozen into the moon's rocky structure, as there is in many asteroids, so you might see a bit of ice. Perhaps Phobos's best attraction would be the view it provides. Phobos orbits Mars three times every Martian day, and it's close enough that you'd see a lot of detail as you sailed over the Red Planet's surface. Sadly, Phobos won't be an attractive tourism destination for all that much longer. Its orbit is decaying, and in about 50 million years it will fall and make a large dent in the Martian surface.

Jupiter is the next planet out. It demonstrates that, while moons are rare in the inner Solar System, they're the norm for the giant planets. Jupiter has more than sixty moons, though most are little more than large rocks. The largest four, however, are very large – large enough that they could be discovered by Galileo when he trained his telescope on the planet in 1610. Europa is the smallest, with a diameter of about 3,100 kilometres (just a little smaller than our moon), but it may be the most interesting moon in the Solar System as it is the most likely place in the Solar System to find simple life. The first close-up images of Europa were taken in 1979 by the *Voyager* missions, but they were followed up by the *Galileo* probe that arrived at Jupiter in 1995 and took detailed pictures of the Europan surface. The pictures revealed something unprecedented in our solar system. Europa's surface, unlike that of most moons, is quite smooth. Its only surface features are complex cracks and ridges that are at most a few hundred metres high. The surface of Europa seems to be an icy shell a few kilometres thick floating on an ocean of liquid water that could be as much as 50 kilometres deep. The cracks and ridges are thought to be pressure ridges generated when large ice plates rub against each other, much the way pack ice behaves in Earth's Arctic ocean. Europa's ocean is probably kept liquid by heat generated by the strong tides Jupiter exerts. The tides squeeze and knead Europa's rocky core, generating heat and possibly even some underwater volcanic activity. This heat keeps Europa's ocean liquid and might also provide the energy for life. Life on Earth exists (and may well have originated) at hydrothermal vents deep in the oceans, where chemicals provide energy to feed organisms at the start of the food chain. It's possible a similar environment exists on Europa. NASA scientists take this idea so seriously that they crashed the *Galileo* probe into Jupiter in 2003 when its mission was over, rather than run the risk of its hitting Europa and contaminating the environment. A visit to Europa could involve some undersea eco-tourism, if we can only figure out how to get through kilometres of ice into the ocean and do so without leaving any trace of us behind.

Jupiter's other large moons are also interesting, if not quite as attractive as Europa. Callisto and Ganymede are the heavyweights at about 5,000 kilometres in diameter. Ganymede is slightly bigger than Callisto and is, in fact, the largest moon in the Solar System. The *Galileo* spacecraft's scans and images revealed much about both planets, which had been thought to be dead and boring moons. It turns out that both have heavily cratered surfaces, the result of a long history of bombardment by asteroids, comets, and meteorites. It also seems as if Ganymede, and possibly Callisto, were geologically active in the past, with signs of large volcanoes and geological faults indicating that plate tectonics may have been at work. What's fascinating about both moons is that their cratered surfaces are probably a mixture of rock and ice, quite unlike our moon. It's thought that both planets might have salty liquid water oceans under a thick crust (perhaps more than 100 kilometres) of frozen rock and ice. Since they're farther away from Jupiter than Europa, they don't experience the same tidal friction and so aren't as warm as Europa. This would explain why their oceans aren't as large or as close to the surface. The water they hold deep within them makes it at least faintly possible that they could host life, but it will be far more difficult to drill down to discover it than it would be on Europa. So Ganymede and Callisto may never be as popular among tourists as their smaller, frozen cousin. They might also have trouble competing with the spectacular sights of Jupiter's most active moon, Io.

Io is Jupiter's third-largest moon. Unlike its icy cousins, Io bristles with spectacular, active volcanoes, which spew lava, gas, and hot dust into space. Io lies closer to Jupiter than the other large moons and so experiences the strongest gravitational tides. These are incredibly powerful. Io's surface is solid rock, not liquid water (where we see tides on Earth), but Io's tides can raise the solid surface of the moon by 100 metres. In effect, Jupiter's gravity kneads the planet like a child's hands knead Silly Putty, softening and melting its rocky interior. Io's surface has no impact craters, as they're constantly destroyed or buried by flows of molten lava. Some debris from Io's volcanoes is stolen by Jupiter's powerful magnetic field, which rips away material ejected into space before it can fall back to the moon and ionizes it. The debris follows Jupiter's magnetic field and feeds huge auroras that illuminate the upper atmosphere of the giant planet. Io would certainly offer many spectacular sights to a tourist, but given all those volcanoes and lava, you'd want to watch where you step.

Saturn, of course, is most famous for its rings, but its moons have a lot to offer as well. Saturn, like Jupiter, may have more than sixty moons (some of the more recently discovered ones are unconfirmed), though many of them are quite small. They all orbit beyond the rings, and the closer moons are thought to play an important role in "shepherding" the rings – keeping them well defined by their gravitational action.

Saturn's moons show a striking variety. Some are rocky, others are icy. Some are heavily cratered and probably billions of years old, and some seem new and may be fragments of older bodies destroyed in collisions. Many of them have been known for centuries (Titan was the first to be discovered in 1655), and others have been discovered only since the *Cassini* spacecraft arrived at Saturn in 2004 and began to take pictures of its rings and moons. Several of the moons have interesting features. Rhea, Tethys, and Dione, for example, are airless ice-balls ranging from 1,000 to 1,500 kilometres in diameter. Mimas is about 400 kilometres across and orbits relatively close to the rings (and is thought to be a shepherd moon). It is marked by a huge crater nearly one-third the diameter of the entire moon. Iapetus is larger, nearly 1,500 kilometres across and is peculiarly two-faced: the front half of Iapetus (the part that leads its orbit) is a dingy black, but the back half is snow white.

Saturn's most remarkable moon by far, however, is its largest one, Titan. Titan is the second-largest moon in the Solar System (after Jupiter's moon Ganymede) and is bigger than the planet

Mercury. It's also the only moon with a thick atmosphere – denser than Earth's, though it's far from breathable since it's made up mostly of nitrogen and methane. Titan was long one of the most mysterious places in the Solar System because its atmosphere is opaque, and through telescopes it appeared to be just a brown and hazy ball. However, the *Cassini* space probe has opened a new window on Titan, first by doing radar and infrared mapping of its surface, and then by dropping the *Huygens* probe into the atmosphere for a descent to the surface. The probe showed us a world that combines the strange and the familiar. Radar images had earlier revealed deserts and dunes and huge lakes, which because of the temperature on Titan – minus 178 degrees Celsius – couldn't be lakes of water. In fact, they're made of liquid methane. Pictures from the *Huygens* probe showed what look like river valleys and deltas from the flow of methane across the moon's surface. The deserts and rocks aren't made of sand but of ice that will never melt. Titan could well experience weather – methane rain, for example. Astronomers think it also may have methane ice volcanoes, as the methane in the atmosphere breaks down over time and must be replenished from layers deeper inside the moon. The thing that might discourage tourists most about Titan is the smog. As the methane breaks down in the atmosphere under the influence of light, many of the same chemicals found on a smoggy day in a large city are produced, things like benzene, ethane, and other hydrocarbons. Not that you'd be breathing them, of course, since you'd have to be safely isolated from the native atmosphere during your stay on Titan. In other words, it would be a fascinating place to visit, but you wouldn't want to live there. In fact, no life that we can imagine could live on Titan's frozen surface.

We know much less about the many moons orbiting Uranus and Neptune because, apart from brief fly-bys by the swift-travelling *Voyager* spacecraft, they haven't been explored. There are many of them (Uranus has more than two dozen, Neptune more than a dozen), and most are small compared to the large moons of Jupiter and Saturn. The great exception is Neptune's moon Triton, which is 2,700 kilometres in diameter, and was discovered in 1846 almost immediately after Neptune itself was discovered. Triton's size isn't the only strange thing about it. Almost everything is a little odd. It has an unusual retrograde path, which means that it orbits around Neptune in the opposite direction to the planet's rotation. It's also a relatively rocky body compared to other large moons. This has led astronomers to suggest that Triton didn't form with Neptune, but is one of the largest gravitationally captured objects in the Solar System. Triton might be an appropriate destination for those who like winter vacations, as its surface temperature is only 38 degrees above absolute zero, or minus 235 degrees Celsius, which makes it the coldest surface in the Solar System. Nice, perhaps, for a little winter camping.

The last moon on our itinerary is Charon, though its classification as a moon is now in doubt. Pluto, as you may have heard, is no longer considered a planet, and so Charon, which circles around it, may no longer be a moon. In fact, Charon's status has always been a little dicey. Charon and Pluto are very close in size – Charon's diameter is a little more than half that of Pluto's – so it's more accurate to say that they orbit each other than to say that Charon orbits Pluto. The two are also in gravitational lock and always present the same face to one another. This makes them more of a binary system than a planet–moon combination. There's little more we can say about Charon. No space probes have come close to either Charon or Pluto, so we have no good pictures of their surfaces, though Charon's seems to be covered with water ice, rather than the methane ice that is thought to cover Pluto. One interesting detail about the couple is that the start of their relationship may have something in common with Earth and our moon. Our moon was formed as a result of a massive collision of some large body with the Earth. It's thought that Charon might be the result of a similar impact when something large hit Pluto. It's always the case with travelling. The farther you go, the more things remind you of home.

Is there life in
our solar system?

As far as we know, Earth is the only planet in the Universe that harbours life. We haven't looked many other places, of course. Apart from that detail, we also don't know much about what processes are involved in the origins of life. So we don't have a very good sense of how likely it is for life to have arisen elsewhere. Life on Earth could be the result of an improbable chemical accident – the result of a lucky roll of the dice. It could, on the other hand, be part of a pattern that is commonly repeated on planets across our galaxy and throughout the Universe. Of course, it could be something in between. The Universe is so unimaginably huge that many people think the chances that life hasn't appeared elsewhere are vanishingly small. There's considerable speculation today that we might not have to go all that far to find extraterrestrial life – it might even exist in our solar system.

If this speculation is correct, however, the life we'd find is not likely to be super-intelligent grey-skinned aliens armed with probes and implants descending from the sky in flying saucers to capture unwary Earthlings. That kind of life is not at all likely (and besides, we don't need to look for them – they'll find us). But even simple life is still worth looking for. It would be almost as exciting to discover whether a simple life form, like bacteria, has found a toehold (okay, a flagellum-hold) somewhere in our solar system. This kind of extraterrestrial life would be a strong indicator of whether life in the Universe is common or rare.

This idea has gained considerably more purchase in the scientific community in the last few years as we've discovered just how robust life is, and how it can survive and thrive in the most extreme and hostile environments on Earth. Scientists have dubbed the creatures that tolerate these environments "extremophiles" – lovers of extreme conditions. One such place is the hot vents of the deep oceans. At these sites there is no sunlight and extreme pressure. After circulating deep in the Earth, water boils out of the ground through fissures in the rock, picking up heat from the magma under the thin oceanic crust. Because of the extreme pressure, the temperature at these vents can easily exceed the boiling point of water on the surface. The hot water dissolves minerals, so the erupting water is rich in minerals and is often acidic as well. Nevertheless, some bacteria thrive in these environments, living on the energy in the chemicals in the water, and these bacteria form the base of a food chain that supports communities of exotic simple animals, including worms and crustaceans.

On the other extreme, we've found life in the Arctic and Antarctic that can survive at temperatures well below freezing.

These bacteria live in briny water and can survive being frozen, thriving again once thawed. We've found bacteria living in rock kilometres underground, surviving on the hydrogen that percolates through the rock. We've found bacteria that can survive radiation levels a thousand times that which would kill a human. Finally, in a discovery still disputed by scientists, an American group claimed in 2001 to have found bacterial spores that have survived as long as 250 million years sealed in salt crystals from an ancient sea and that, when dug up, grew again.

Clearly, simple life can tolerate a range of conditions far wider than we fragile humans can handle. As a result, there's considerable curiosity among astrobiologists about where in the Solar System we might find suitable environments for simple organisms that can tolerate harsh conditions. It turns out there are a few likely places, a few less likely places, and a great number of places that are entirely ruled out. We set several preconditions when looking for these places. The first is that life is likely to operate according to the same chemical rules that life on Earth does, which is to say it will use organic carbon chemistry using water, and it will need to extract energy from its environment. Well, the Solar System is full of simple carbon compounds, including amino acids – known as the building blocks of life – and such hydrocarbons as methane. Energy is available from the Sun, but in the more extreme environments it's more likely that life would harvest chemical energy – again perhaps from hydrocarbons or the kinds of chemicals produced in volcanoes – like those consumed by sulphur-eating bacteria living on Earth.

The second crucial precondition is water, which is a much rarer commodity in our solar system. To date we've found relatively few places where water could exist in liquid form and wouldn't be frozen into ice. The short list includes only one planet – Mars – and a few of the moons around the giant planets. The inner planets – Mercury and Venus – are simply too hot. The outer ones – Jupiter, Saturn, Neptune, and Uranus – are caustic, high-pressure gas giants that couldn't sustain liquid water or life. There is such a thing as too extreme, even for extremophiles.

In recent years, there's been considerable interest in looking for life on Mars, partly because it's the best candidate for life we've found so far, and partly because at the moment it's the most accessible planet for exploration. With orbiting satellites and multiple robot landers, we've arguably done a better job mapping and exploring the surface of Mars than we have of the Moon, despite the fact that humans have never been there. Mars is such a tempting site because it seems, in its ancient history, to have had an environment much like Earth's, with oceans and a thicker atmosphere than exists today. The landscape has features that appear to have been carved by running water, and from what we know of planetary formation, Mars might have been quite a friendly place – as friendly as early Earth – for simple life. Unfortunately Mars's window of habitability closed after probably only a few hundred million years, when it became the dry, frozen, nearly airless planet that we see today.

However, if conditions were once benign enough for life to get started on Mars, it might have persisted in tiny pockets that robotic explorers or future human visitors might find. We think Mars's surface can no longer harbour life, but it may persist below the surface. The key is whether there is any liquid water underground. Mars doesn't seem to be warm enough to maintain geological activity – it has no active volcanoes – but it may still be quite warm deep under the surface, and liquid may exist there. Orbiters like the *Mars Global Surveyor*, which arrived in Martian orbit in 1997, have been watching the planet's surface for several years now, and several sites have been found that look like the result of recent geological activity. In 2006, there was great excitement when the satellite produced pictures of what scientists think could be a new gully produced by an outburst from an underground spring. The gully wasn't there in a picture taken in 2001. If it could be confirmed that the trace on the surface the orbiter saw was water, then we'd be more confident that Mars could sustain life.

Robot explorers may be able to find these signs, but even the leaders of the robotic missions assert that humans would do a better job, which is one of the key motivations for a mission to Mars. The paradox of such a mission is that humans would bring their own bacteria to Mars, which could contaminate the Martian ecosystem, possibly jeopardizing or even extinguishing any life there.

Other candidates for life include Europa, a moon of Jupiter, and Enceladus, which orbits Saturn. Both of these moons seem to be ice-balls. Europa is a possible site for all the reasons explained earlier, in the answer to Question 18. Enceladus, a tiny moon of Saturn, is less than 500 kilometres in diameter, and when *Voyager* flew by it in 1981, scientists were surprised by its smooth surface, suggesting that it, like Europa, has had some kind of surface renewal in the recent past. They were even more stunned in 2005 when the *Cassini* probe flew by the little moon and took photographs of deep cracks in the surface and plumes of water vapour shooting out. An analysis of the water revealed that it contains carbon dioxide and such hydrocarbon molecules as ethane and propane. Enceladus, like Europa, seems to be heated internally by tidal forces from Saturn's gravity, and this is likely what is keeping water liquid within it. Whether life could arise on a tiny moon like this is an open question, but what we think are the necessities are there.

Space exploration can only reveal so much, and we may find out far more about the chances of finding extraterrestrial life from our work in biology labs than we will from visiting distant moons. While places like Europa and Enceladus have what we think are the ingredients for life, we know little about how those ingredients have to come together to create self-replicating organisms. It's all very well to have the ingredients, but unless you've got the recipe, you'll never have cake.

WHY ISN'T PLUTO A PLANET?

IN 2006 AT A MEETING of the International Astronomical Union, Pluto, formerly the ninth planet, was officially reduced in status to a dwarf planet. It is now a planetary wannabe – a minor-league planet denied the glory that real planets have. Some people were outraged. Some people celebrated. Some people shrugged their shoulders and wondered what all the fuss was about.

This was the conclusion to a tussle between two groups of astronomers we'll call the rationalists and the sentimentalists. The rationalists insisted that they needed a solid, consistent definition of what a planet is in order to avoid confusion in astronomy. Inclusion of Pluto, they claimed, made the category of planet useless because Pluto was, in important ways, not like the other things we call planets, and too much like things we don't want to call planets. Therefore, Pluto was out. The sentimentalists, on the other hand, remembered Pluto fondly, not just as a cartoon dog, but as one of the canonical nine planets that we all learned about in school. Losing Pluto as a planet would be like losing Antarctica as a continent; we'd somehow be diminished by it. The maps and textbooks would be wrong, and a generation of children confused. The rationalists seem to have won this round, but it may not be over. Discontent still simmers on this issue, and the sentimentalists may rise again to reinstate their favourite.

What this really illustrates is just what we know, what we don't know, and what we've learned about planets in a relatively short time. Pluto's demotion is an illustration that the Universe – or at least our knowledge of it – is quickly changing as we discover more of its details. Our picture of our solar system, our galaxy, even our cosmos, is constantly being altered as we see farther and in more detail.

It's not as if Pluto had been a planet for all that long. Nor is it the first planet to be demoted. Many of the planets of our solar system have been known since antiquity because they can be seen with the naked eye. They were known to the ancients as the "wandering stars," because unlike the real or "fixed" stars they moved about the sky. They were Mercury, Venus, Mars, Saturn, and Jupiter. Add Earth for the six original planets (though of course the ancients didn't think all these bodies were the same). Uranus was the first planet to be discovered by telescope in 1781, and the discovery of Ceres followed in 1801. Not familiar with the planet Ceres? No surprise there, as it was the first "planet" to be demoted. An Italian priest and astronomer named Giuseppe Piazzi spotted it and claimed it as a planet, but not long afterward the discovery of several similar bodies made astronomers rethink its planetary status. Ceres was then reclassified as just one of the largest of the asteroids in the asteroid belt, which set the precedent of kicking astronomical bodies out of the planetary club. The first sighting of Neptune followed in 1846, but then there was a long dry period. Pluto wasn't discovered until 1930, when astronomer Clyde Tombaugh, using quite a small telescope by modern standards, teased a tiny point of light out of the stellar background in a remarkable bit of observation. So Pluto is a planet-come-lately, and while it's true that most of us learned that Pluto was a planet in our science textbooks, there is a (shrinking) population of people for whom the idea of nine planets is new-fangled nonsense. You've heard of the famous piece of classical music inspired by our solar system, *The Planets* by Gustav Holst? Well, check the record jacket. Guess what? No Pluto.

While Pluto entered the planetary club in 1930 at its discovery, it's been an inconstant member. Pluto is small and very dim. Even on our most powerful ground-based telescopes it appears as nothing more than a few pixels – a tiny spot of light – on their digital pictures. As a result, we've constantly been revising our notion of what Pluto actually is. When it was first discovered, it was thought to be as large as Earth. In the 1960s, it was realized it was smaller, perhaps only half the size of Earth. In 1978, we learned that Pluto had a large moon – Charon – and that all this time astronomers had been including Charon's brightness with Pluto's in measuring its size. That made Pluto smaller still. Today we think Pluto is about 2,300 kilometres in diameter – only about two-thirds the size of our moon. In fact, there are seven moons in the Solar System bigger than Pluto, and it's the smallest planet by far (Mercury is twice as big).

There's nothing wrong with being small, of course. But other things have come to light that make Pluto's planetary status awkward. The first of these was hinted at by Pluto's irregular orbit. The orbits of the rest of the Solar System's planets lie, more or less, in a single plane around the Sun. Pluto, though, has a massively tilted and extremely elliptical orbit. Its orbit actually cuts inside Neptune's. A Plutonian year is 248 Earth years, and for twenty of those years it's not the most distant planet – Neptune is. This occurred most recently between 1979 and 1999. Watch for it to happen again in the year 2227. Now an irregular orbit is also nothing to be ashamed of, but it is indicative of a bigger issue – the company Pluto keeps.

In the early 1990s, astronomers began to discover other objects in the distant reaches of the Solar System, most just a little farther out than Pluto. These were also tiny dim things like Pluto, many of them in irregular orbits, most of them smaller than Pluto. Soon the numbers of these things reached the dozens, then hundreds. Astronomers now think there are at least 70,000 of these attractively named Kuiper Belt Objects (KBOs) that are larger than 100 kilometres across. They whirl around the Solar System in orbits that range from near Neptune (about 30 times the distance from Earth to the Sun) out to vast distances (perhaps 1,000 times that distance). More significantly, beginning around 2000, astronomers began to discover objects approaching Pluto's size. We now know of nearly a dozen objects bigger than 1,000 kilometres across. The straw that broke the camel's back was the discovery in 2005 of the object later named Eris (appropriately, the Greek goddess of discord and strife), which according to observations made in 2007 is actually a little larger than Pluto. Eris is a true wild card. Its orbit is extremely elliptical, as it cuts inside Pluto's orbit at its closest approach and then goes out more than twice as far as Pluto's orbit at its most distant, taking more than 550 years to circle the Sun. Its orbit is also tilted nearly 45 degrees to the plane of the rest of the Solar System.

What all these discoveries meant was that if we were going to call Pluto a planet, then there was no good reason for excluding Eris from planetary status, and perhaps no good reason for excluding a lot of the other large KBOs as well. The real problem was that the International Astronomical Union (IAU) didn't have a definition of a planet. Traditionally, the approach astronomers took to identifying planets was that they'd know one if they saw one (it's not a question that comes up all that often). Once the KBO candidates started showing up, the problem of an imprecise definition emerged. Scientists like to have neat categories and definitions, since the precision minimizes confusion in their discussions. It also had implications for the textbooks of the future. Just how many planets was our solar system going to have? After all, if you started letting in things like Eris, and other KBOs, what about Ceres, or the large moons of Jupiter?

The IAU started looking at various definitions of what a planet is and voted on them at a meeting in 2006. They agreed on a couple of key points. One is that a planet has to be large enough that its gravity would force it into the shape of a sphere. This knocked out the oddly shaped small fry. The second characteristic is that the object has to be in orbit around a star and can't be a satellite of another planet. This knocked out the large moons. Finally, and this was the fatal blow for Pluto, a planet has to have "cleared the neighbourhood around its orbit," which is to say that a planet can't be part of a club of objects peacefully co-existing as orbiting neighbours. It had to be the ruler of its orbital domain, having either captured or gravitationally ejected everything nearby. If it hadn't cleared the neighbourhood, as far as the IAU was concerned, it wasn't a planet. Since Pluto is part of a family of Kuiper Belt Objects, and Ceres was part of the asteroid belt, they both became dwarf planets in August 2006. Calling Pluto a planet is now punishable by the IAU's highest sanction, which is – well, nothing really. It's just not done any more, and astronomy geeks will look down their telescopes at you and sneer. Best avoided.

This might seem like a tempest in a teapot, but the definitions set in 2006 will be useful in the future. There are doubtless more large KBOs out there waiting to be discovered, and now we'll know what to call them. Perhaps more importantly, as we develop the technology to look at other stars and the objects that might be orbiting them, we'll have a clear and accurate way to describe them. In a way, it also helps us to build a more accurate picture of our solar system – especially of the mysterious and remote depths beyond Neptune. The old picture was of the neat orbital shells of eight planets and the not-so-neat orbit of the ninth. Now we know that the Solar System's outer reaches are more interesting and more crowded than we'd ever imagined. It's a new space frontier to explore.

Is there a tenth, I mean ninth, planet?

Now that Pluto has been demoted to the minor leagues, you might be wondering if, when it comes to planets in our solar system, eight is enough. Is it possible that we'll find another real planet, perhaps in the distant outer reaches, beyond the Kuiper Belt? Perhaps there is something even bigger than a planet – a dim distant companion star for our sun. This is an idea that's transfixed some astronomers at least since the discovery of Pluto. It's also an idea that's waxed and waned in popularity as we've explored the Solar System, and for some it has come to have a vaguely disreputable aura. But we don't know for sure if we've discovered every planet, and there remains some tantalizing evidence that there could be something out there still waiting to be found.

The idea that there might be another planet beyond the orbit of Neptune dates back to the late nineteenth century. Observations of Neptune and Uranus at the time hinted that there was something strange about their orbits. Their motions through the sky when initially measured seemed to be slightly unpredictable, as if they were being affected by the gravitational influence of another large, as yet undiscovered planet. This suggestion wasn't far-fetched. The existence of Neptune itself had been predicted by observations of the unexplained motions of Uranus before Neptune was actually discovered in 1846.

One devotee of this idea was Percival Lowell, the American astronomer famous for his observations of Mars and his conviction that he could see canals signifying intelligent life on the Red Planet. Apart from his interest in Mars, he was also fascinated by the search for the ninth planet, which he called "Planet X." Then, fourteen years after Lowell's death, Pluto was discovered at the observatory named in Lowell's honour. But Pluto wasn't Lowell's Planet X. It wasn't large enough for its gravity to influence the orbits of the giant outer planets to the degree that astronomers thought they'd observed. According to their calculations, Planet X would have to be quite large – at least six times the size of Earth.

The search for Planet X continued for nearly fifty years. In time, many astronomers came to believe that no such planet existed. They suggested that the unexplained motion of Uranus and Neptune was just an error in observation. The end for Planet X came as space probes – *Pioneer 10* and *11* and then *Voyager 1* and *2* – approached the outer planets. None of these probes encountered any gravitational disturbance (which would have affected their trajectory) that would correspond to the presence of a large planet. They also were able to more closely examine Uranus and Neptune, and that close examination showed that there was no unexplained influence on their orbits. The unexplained motions simply disappeared once astronomers recalculated their measurements. It had all been a mathematical error based on inaccurate estimates of the masses of the outer planets.

Instead of a giant Planet X, the early twentieth century brought us many new dwarf planets. The Kuiper Belt – the vast space beyond Pluto – turned out to be full of comets, asteroids, and many bodies as large as and at least one even larger than Pluto. This is why Pluto was demoted from planetary status. If Pluto was a planet, so were these other bodies, so we didn't just have Planet X, we had planetlets Q, R, S, T, W, Y, Z, and probably many more. None of these dwarf planets is bigger than our moon, though, and their influence on other objects in the Solar System is negligible.

There are, however, hints that perhaps something else is out there, even farther away. This would not an ordinary planet at some reasonable distance beyond Pluto, but something more exotic and perhaps more interesting.

One such hint is in the form of a phenomenon known as the Kuiper Cliff. The Kuiper Belt consists of small, icy bodies of various sizes orbiting the Sun at about the orbit of Pluto – from thirty times Earth's distance from the Sun (30 AU, or Astronomical Units) out to about 50 AU. At 50 AU, however, there seems to be a drop-off – there isn't much out there beyond that. That's the Kuiper Cliff. There are many potential explanations for this, but one intriguing one is that a large planet orbiting well beyond 50 AU has cleared out that part of the Solar System by capturing or gravitationally deflecting those objects within its influence. This isn't the only possible explanation – it's probably not even the most likely one. The Kuiper Belt's motions are complicated and we don't understand entirely even how it has been shaped by the movements of the giant outer planets, Uranus and Neptune, since the Solar System formed four and a half billion years ago.

One strong argument against a Kuiper Cliff planet is that it hasn't been seen. Astronomers have discovered much smaller objects at extreme distances, and fairly comprehensive sky surveys have looked for just such kinds of bodies without any sign of a large planet out there. On the other hand, the planet would be very dim and would move very slowly, making it quite difficult to see. There is a faint possibility that a faint planet might have been missed.

More exotic than a planet is the remote possibility that our solar system has not one, but two stars, and, if so, the second one could be a killer. This is the Nemesis hypothesis, the subject of wide-ranging speculation on the Internet – much of it mystical and fantastic, and some of it just plain strange. Nevertheless, the possibility is being investigated by a small group of astronomers.

The root of the Nemesis hypothesis is a controversial claim that there may be a reason for the timing of the extinctions we've suffered on Earth. There seems to be at least a low-level extinc tion event about every 30 million years. Extinctions can be caused by all sorts of things, including disease, climate change, tectonic upheavals, and asteroid and comet strikes. Supporters of the Nemesis hypothesis suggest that these causes wouldn't make extinctions occur with such regularity. An astronomical explana-tion solves this puzzle.

The theory's supporters suggest that if the Sun had a small, dim companion star – a red or brown dwarf still significantly larger than a planet – it could cause such events. The star would be in a highly elliptical orbit far beyond even the Kuiper Belt. At its most distant it would be as far as 90,000 AU from the Sun. Every 30 million years or so, however, it would make a close approach to only 20,000 AU, passing near the distant cometary belt, called the Oort Cloud, that forms a halo around the Solar System. In doing so, comets would be captured by its gravity, and some might be flung toward the inner Solar System. Some might have hit Earth, causing mass extinctions.

The idea that the Sun has a companion star is not that radical. Many stars do. Up to half of the stars we see in the sky may be part of binary systems that consist of one large bright star, like our sun, and one small, dim companion, like Nemesis would be. It's also quite possible that a sun like Nemesis could remain undiscovered for so many years. Red and brown dwarfs are very difficult to observe. There are thousands in our stellar neighbourhood, and it's very difficult to determine their distance and size. They move slowly, and stellar motion is the method most often used to determine the orbits of distant objects. It's possible, then, that if Nemesis exists we've already seen it but haven't realized what it is.

It is, however, extremely unlikely. Using data from infrared telescopes, all Nemesis candidates examined so far have been ruled out. The periodic extinctions may have to await another explanation. Mind you, proving that something isn't there is much more difficult than proving that it is. By the way, in case you're worried that Nemesis might be dropping a comet on our head as we speak, don't fret. It isn't expected to come close for another 25 million years or so.

Where does the
Solar System end?

Grade-school science books and encyclopedias generally include a neat diagram of the Solar System showing the Sun in the middle, with the planets neatly spaced out all the way to Pluto. There are several things wrong with this picture. The first is the scale. A bigger problem, though, is that the Solar System doesn't end at Pluto. There is a great deal of Solar System out there well past Pluto, and a lot of it is very interesting – if cold and dark. There are planets (or at least dwarf planets, similar to Pluto), comets, and asteroids in uncountable numbers, and then there are the more unsubstantial things like magnetic fields and the halo, which surround our solar system like a cocoon as it travels through interstellar space. Showing a model of our solar system that includes only the planets is like drawing a picture of a human that includes only the bones. You can get an idea of what it looks like, but there's a lot you're missing.

Let's look at the scale first. The Solar System is quite large, even if we just confine it to the planets. A useful demonstration is to take some common objects and lay them out to get a sense of the distances. Make the Sun a basketball. Drop that on the goal line of a football field. Since it's football, we'll talk in yards, but you could say metres as well – it's close enough for our approximation. Now walk out to the 10-yard line and drop a poppy seed there – that's Mercury. You can drop sesame seeds at the 20-, the 27-, and the 40-yard lines for Venus, Earth, and Mars. Jupiter and Saturn are a respectable size – use a golf ball (though that's a little big). Unfortunately you've run out of football field, you'll need another one, make that two, because Jupiter is at 140 yards, and Saturn at 260 yards. You're on your fifth football field for Uranus (a hazelnut will do) at 520 yards, on your eighth for Neptune (another hazelnut) for 820 yards, and nearly on your eleventh for poor, demoted Pluto, for which you'll need a grain of salt. That basketball – the Sun – doesn't look that big from here, does it? Scale this up to the real size of the Solar System, and at Pluto's average distance from the Sun you're about 5.8 billion kilometres away from our star.

That's the Solar System we're familiar with, but it goes out much farther than that. Beyond the planets is the Kuiper Belt, the Heliopause, and beyond them the mysterious Oort Cloud, which may range beyond what we would call the edge of the Solar System. That takes us out more than 14 billion kilometres from the Sun. In our scale model, the Solar System ends fully 3 kilometres – roughly thirty football fields – from that basketball on the end zone of the field. That's probably why those textbook diagrams aren't to scale. Even if you shrunk the Sun to the size of a sesame seed, you'd need a forty-page fold-out to put Pluto on it to scale, and about a hundred for the Heliopause.

We've only just begun to sort out what's going on in all that space beyond Pluto. We have spacecraft that have now travelled that far – *Pioneer* and the two *Voyager* probes are getting toward the edge of the Solar System, but they weren't designed to explore the outer reaches. The *New Horizons* probe, launched in 2006 and on its way to the outer Solar System now, is the first mission specifically designed to study Pluto and beyond. Our telescopes, including space-based telescopes like Hubble, have improved considerably, so that we can actually see a bit more of the outer Solar System. It is still a difficult task, as most of what is out there are relatively tiny, poorly illuminated bodies.

The first layer of the outer Solar System – the Kuiper Belt – has been the biggest surprise so far. You can't really call it the next thing past Pluto, as it's clear now that Pluto is part of the Kuiper Belt. (You didn't skip the answer to Question 21, did you?) The Kuiper Belt is considered to occupy the space between Neptune's orbit at 30 AU out to 50 AU. It's a large volume of space – much larger than the area inside Neptune's orbit where the planets dwell. Since the early 1990s we've known that this part of the Solar System contains billions of objects and a huge amount of material. There are an estimated seventy thousand objects bigger than 100 kilometres across, and there is at least one other object bigger than Pluto – the dwarf planet Eris – and likely more that we haven't discovered yet. It's thought that the total mass of everything out there might be as large as the mass of Jupiter.

Pluto, again, gives a clue to what all these Kuiper Belt Objects are like. Pluto is an unimaginably cold ball consisting (according to best guesses) of a rocky core surrounded by water ice with a crust of frozen methane and nitrogen. The frozen methane is reddish-brown where sunlight has broken it down, but any new impact craters and the debris around them is bright white where fresh ice shines through. Pluto is, in some ways, a lot like a giant comet, and most of the Kuiper Belt Objects are very likely similar to Pluto, though some may be brighter and some darker, depending on how much of their surface has been disturbed. Many comets come from the Kuiper Belt, while others come from farther out in the Oort Cloud. What we're looking at when we see KBOs is the same sort of material that was present in the earliest days of the Solar System. The Kuiper Belt might even be material from the beginning of the Solar System that didn't form into true planets. Instead it gathered into many smaller objects, and these are moved around and constrained by the gravity of Neptune. Think of the Kuiper Belt as a thin ring around the Solar System, not too different from Saturn's rings.

We've extended the Solar System nearly twice as far as Pluto with the Kuiper Belt. There are objects that are nominally part of our solar system even farther out. They're part of the mysterious Oort Cloud, which is interesting largely because it has never been seen. Jan Oort, a Dutch astronomer, first realized the possibility of the cloud in 1950. He'd been studying comets with orbits that take hundreds of years to complete and realized that given these orbits they must have come from a tremendous distance. Comets tend to have very elliptical orbits, and Oort calculated that, at their farthest from the Sun, they could reach 50,000 AUs – swinging a tremendous distance out before the Sun's gravity reasserted itself and dragged them back into the inner Solar System. Oort also recognized that comets seem to come from all directions. The picture he came up with is of a vast cloud or shell of comets orbiting at tremendous distances from the Sun – possibly a trillion of them. A tiny few are disturbed in their long quiet orbits by the passing of nearby stars, and this causes them to fall in toward the Sun, becoming the spectacular comets we occasionally get to see. Once we can see them, though, they're doomed. Their bright tails represent their destruction as they lose material on their approach toward the Sun, and they all eventually evaporate away to nothing. It's only the comets that have been thrown out of the Oort Cloud that we ever see. No telescope yet invented has the power to see that far out into space and show us an Oort Cloud object in its native habitat.

Unlike the Kuiper Belt, the Oort Cloud is not considered to be a remnant of the formation of the early Solar System. Although they are farther away than the Kuiper Belt, the objects in the cloud are thought to have begun farther in. These comets probably formed in the vicinity of Neptune and Uranus in the early Solar System, but the gravity of these massive planets simply flung them away, into the distant reaches of the Solar System. Not all of them were sent into the Oort Cloud, of course. Some were probably thrown into the inner Solar System, hitting the planets or dropping into the Sun. Some were likely thrown out of the Solar System entirely – ejected at a speed that allowed them to escape the Sun's gravity entirely and drift off into inter-stellar space. Nevertheless, a large number seem to be still with us, at extreme distances from the Sun, just close enough to be still under the influence of its gravity.

While they're in orbit around the Sun, many Oort Cloud objects may not be in the Solar System at all. It depends on where you draw the line, and increasingly astronomers draw that line not where the Sun's gravity becomes insignificant, but at the point where the solar wind peters out, known as the Heliopause. The solar wind is a constant stream of particles emitted by the Sun at speeds well in excess of a million kilometres an hour. It pushes out, creating a kind of bubble of hot gas around the Solar System. This bubble is known as the Heliosphere, and its edge is the Heliopause – the point at which it has slowed and become thin enough that the pressure of the tiny number of particles in interstellar space pushing back on the solar wind balances it out. Think of the Heliosphere like the Solar System's bow wave as it sails through interstellar space. Beyond the Heliosphere you really are between the stars, in the medium through which our solar system is quietly drifting. The Heliosphere isn't entirely accurately named, because it's not a sphere. Since the Solar System is moving through space, there's pressure on the solar wind, which distorts the "sphere." It is compacted in the direction in which the Solar System is moving and expanded behind, in a teardrop shape. If the front edge of the Heliosphere is the bow wave of the good ship *Solar System*, behind us we leave a solar wake. The bow wave thrown up by the Heliosphere acts like a force field for the Solar System. It deflects a large proportion of the intergalactic cosmic radiation flying at us from space.

We've recently begun to explore the edge of the Solar System. The *Voyager 1* spacecraft is thought to have entered the region of the Heliopause in 2005 at a distance of about 90 AU. Unfortunately, one of its chief instruments for measuring the solar wind is no longer working, so scientists are getting less information about it than they'd like, but they're lucky to be getting anything at all from a thirty-year-old probe. Sometime in the next decade or so, *Voyager* will cross the boundary into interstellar space and thus will become the first human-made object ever to leave our solar system.

How long until we have to
leave the Earth?

THERE COMES A TIME in many homeowners' lives when the decision has to be made: should we stay in this old house, or is it time to move? Houses age. The plumbing and wiring need replacing. The foundation needs work, the roof is starting to leak. The place is drafty, too hot, or too cold. Maybe the family is bigger and needs more room. Maybe the neighbourhood isn't so good any more.

As with houses, so too with planets. Over time – admittedly a large amount of time – the Earth is going to decay and stop being as homey. This might be because of pollution and humans "using up" the planet, but even if we're able to stop our environmental degradation, over the long term, Earth just won't be so nice any more because of several factors we can't control.

For one thing, we're eventually going to lose our maintenance and renovation service. The Earth has been continually evolving since it first formed, thanks largely to the process of continental drift and plate tectonics. These are relatively new ideas. The theory of continental drift was only developed in the early twentieth century, and plate tectonics as an explanation for continental drift didn't come along until the 1960s. Geologists are still trying to understand the details of how this works, but the basic idea is that the surface of the Earth is made up of plates of crust – solid rock – which drift on an interior ocean of semi-molten rock. The drifting of these plates causes them to collide, lifting up new land and burying old. As a result, the Earth's surface is constantly being recycled as new material is introduced into the biosphere, and old material driven deep underground to be melted down. Apart from keeping the surface of the planet rejuvenated, this has had important effects on such things as carbon dioxide levels in the atmosphere. Carbon is taken out of the atmosphere by biological processes and dissolved into the oceans, where it sinks and can be buried, and then taken deep into the interior. It's also reintroduced into the atmosphere by volcanoes. These processes are vital to the biosphere. We have a problem with CO_2 levels being too high now, because of an excess produced by human activity. There is, however, a natural and essential greenhouse effect that we depend on for life on Earth. If not for it, the temperature of the Earth would average about minus 18 degrees Celsius, the oceans would freeze, and the Earth would be a snowball. So the recycling of carbon dioxide by the hot core of the Earth is one of the things that makes life as we know it possible.

Even so, the Earth is slowly cooling. There are several reasons for this. Part of the heat in the Earth is residual warmth from the formation of the planet. The outer shell of the Earth has cooled and solidified, but the inner part retains the latent heat of planetary formation. This heat is slowly radiating away into space. Gravity is also heating the planet, as it "sorts" the elements into layers in the interior – dragging the heavier ones deeper toward the core and creating heat through friction. This will end eventually when the sorting is complete. The biggest source of heat in the planet is the radioactive decay of elements trapped within the planet when it formed. Radioactivity, of course, declines over time as well. So as radioactivity declines, this heating of Earth's core will end, and the continental plates will freeze in place. Our living Earth will become, geologically, a dead planet.

This will have multiple effects. One is that Earth will lose its protective magnetic field, which is generated by the rotation of the core within the planet. That will mean a large increase in the amount of radiation that hits Earth's surface. Radiation is not the biggest worry, though. There will be no more volcanoes and no more new mountains. Carbon dioxide will slowly leach out of the planet's atmosphere, which means the planet will begin to freeze. Both the loss of CO_2 and the cooling trend will devastate plant life, which forms the base of the food chain. A geologically dead Earth will be followed soon after by a biologically dead Earth.

That's a nasty scenario, but you don't have to worry about it, as it will take many billions of years for Earth to cool that much. Long before that happens the planet will be destroyed by the aging and death of the Sun.

Our sun is a common enough star, and by observing similar stars at different ages and developing models of how these stars operate, astronomers have created an evolutionary timeline for our star. What's most important is what goes on at the core of the Sun, where pressure and heat keep the nuclear fusion reaction in the star burning. When the Sun was born, its core was about 75 per cent hydrogen and 25 per cent helium. Nuclear fusion in the core converts hydrogen to helium, and so over time the hydrogen is consumed. Right now, for example, the core of the Sun is down to about 30 per cent hydrogen and 70 per cent helium. In about five billion years, nearly all of the hydrogen in the core of the Sun will have been consumed. At this point the Sun is going to undergo a fairly radical change. The core, now entirely helium, will contract. Fusion of the remaining hydrogen outside the contracting core will cause the outer layers of the star to swell. In a few million years the core will have contracted and heated up enough that the helium in it will begin to undergo nuclear fusion, changing into carbon and oxygen, and the Sun will enter a new age of stability lasting billions of years. This reborn and much larger Sun will have evolved into a different kind of star – a red giant.

The Sun's growth will not be good for the Earth. It's expected that it will swell beyond the Earth's current orbit. Our planet will be incinerated, and as the Sun goes through its next stage, during which it blows off its outer layers, the ashes of the Earth will go with it.

Not all astronomers agree with this scenario. Some think there may be some hope for Earth. As the Sun grows into a red giant, it will also start to lose mass as some of its material is blown off into space. The Sun's mass, however, is generating the gravity that is holding Earth in its current orbit. If the Sun loses enough mass (and the numbers are uncertain here), Earth's orbital diameter will increase and we'll withdraw from the Sun. It's possible that as the Sun grows in size (but shrinks in mass), the Earth will back off to a safe distance and avoid incineration. Unfortunately the next stage in the Sun's evolution, after it blows off its outer layers, is its collapse into a white dwarf – a nearly dead ember. A white dwarf at the heart of our solar system unfortunately won't be bright enough to heat the Earth to habitable temperatures. That means that if the Sun doesn't incinerate the Earth, it will probably freeze.

Again, though, you don't really need to worry about this either, and not just because you won't be around in five billion years when all this is expected to happen. In fact, the Sun will likely end life on Earth long before it incinerates it by boiling and evaporating the oceans and atmosphere.

The Sun, like the Earth, has evolved over time, becoming significantly brighter in the 4.5 billion years since the Earth formed. At the time the Sun formed, it was only about 70 per cent of its current brightness. Two billion years ago, when life was still just a rumour on our planet, it was about 85 per cent as bright. This increase in brightness is going to continue and will likely destroy the biosphere in as little as hundreds of millions of years. In about 300 million years or so, the mean temperature on Earth will increase by at least 5 degrees Celsius, and this will make large parts of the planet uninhabitable. By about a billion years from now, the Sun will be hot enough that the oceans will start to evaporate. Since water vapour is a greenhouse gas, this is a feedback loop that will accelerate, as more water will absorb more heat and accelerate the heating. This won't go on forever, though. Much of the water vapour in the atmosphere will end up escaping into space, and so we will lose the oceans that made life possible. The planet will then cool somewhat, but it will now be a water-scarce desert planet incapable of hosting life.

We may, however, be able to engineer a reprieve from this fate. It is, after all, a long way off, and some astronomers have suggested we find a way to move Earth gradually farther away from the Sun so as to keep its effective brightness the same. The idea involves nudging the orbit of a large asteroid so that it makes frequent close passes to Earth. At each encounter, the gravity of the asteroid would drag the planet a tiny bit farther away from the Sun, slowly increasing the diameter of our orbit. Since the Sun's brightness will also change slowly, these tiny tugs, and the resulting tiny increases in the Earth's orbit, will offset each other perfectly, and life on Earth will be preserved. Using this method we might survive past the 300-million-year deadline the Sun has imposed.

At least until the next disaster. . . .

WHY WON'T WE GO TO THE STARS?

HUMANS ARE EXPLORERS. We've looked at most of the nooks and crannies on Earth (though we've yet to thoroughly explore the oceans). We've also taken baby steps off the Earth. It seems perfectly logical and quite natural to think that at some point we will explore and perhaps settle on other planets, and eventually other solar systems. This idea is a staple of science fiction but there's more to it than that, as the fiction represents what a lot of people believe to be true, especially those involved in our current efforts to explore space. For example, Michael Griffin, the head of NASA since 2005 and a hard-headed engineer, businessman, and administrator, has said he is convinced that humans will colonize the Solar System and, one day, go beyond to the nearby stars.

Sadly, there's very little reason to believe he's right. From everything we know about physics and space, the dream of humans travelling among the stars is probably an empty one. On television, travelling from one solar system or galaxy to another is just a matter of getting your curmudgeonly Scottish engineer to finally get the warp drive on-line. In real life, we're facing challenges that are likely intractable.

The best example? Well, the most efficient chemical rockets we have now are the Space Shuttle main engines. NASA scientists have calculated the amount of rocket fuel it would take to send a Space Shuttle–sized payload accelerated by those engines to the nearest star – Proxima Centauri – travelling fast enough to get there within 900 years. The amount of fuel was 10^{137} kilograms. That's 1 followed by 137 zeros, and that's an amount bigger than the mass of the Universe. This seems absurd, and it is. That's because we generally ludicrously underestimate the distances, time, and energy required for space travel. By the way, we're pretty sure that Proxima Centauri doesn't have a planet worth exploring. We would need to go farther to get to anywhere interesting – much farther.

Let's try to get a sense of what these distances are. Proxima Centauri is about 4 light-years away. We all know that a light-year is the distance that light travels in one year, but it's not a very useful measure for comparison. So in more useful units, that's about 39 trillion kilometres. Actually, those units aren't all that useful, since a trillion is pretty much impossible to imagine. Let's put it in terms of how long it might take to get there at speeds we have already achieved. The fastest that humans have travelled in space to date is roughly 40,000 kilometres per hour. If a spacecraft left Earth's orbit travelling at that speed, it would take more than 100,000 years to get to Proxima Centauri. To put that in context, our entire species is only a little more than 100,000 years old. If we put a family of *Homo erectus* on that spacecraft, they'd have the time to evolve into *Homo sapiens* by the time they got there. . . .

Clearly a 100,000-year journey isn't going to happen, so we need to go faster. The problem is that even using the kind of technologies for space flight we are still only dreaming about, the amount of time and resources it would take for the trip is still, well, astronomical. Let's improve on the example of a chemical rocket. NASA has looked, hypothetically, at the amount of energy and propellant needed if we used more powerful systems for pushing rockets. To set their conditions, they arbitrarily chose nine hundred years as a reasonable amount of time for a one-way trip, and a Space Shuttle–sized payload as what we'd want to send. For a human mission, this is far too long and far too small a payload, but be that as it may.

They looked at a nuclear-fusion–powered rocket, which is something we might figure out how to build in the next fifty years or so, and which would be a hundred times more efficient than a chemical rocket. They calculated you'd still need 10 billion kilograms of propellant, which is the equivalent of a thousand supertankers. So much for nuclear power. The next step up, the most powerful energetic reaction we know of is the matter-antimatter annihilation reaction. Using this, you'd need only about 100,000 kilograms of propellant. This seems more reasonable, and antimatter isn't science fiction. Physicists have created and trapped antimatter, but the amount that has been trapped is measured in picograms – a trillionth of a gram, and it doesn't stay trapped – and antimatter particles tend to repel each other, flying off and exploding violently on contact with normal matter whenever they're concentrated. Clearly we'll have to ramp up to get to the mass production necessary for space travel. We haven't yet even thought about how to design an antimatter fuel tank or engine. There's also a little safety problem. A hundred thousand kilograms of antimatter, exploding on contact with any kind of real matter, is more than enough to destroy our planet.

This has led researchers, unsurprisingly, to the conclusion that using propellant to push a rocket to another star is probably not going to happen. So what are the technologies that don't use propellants? One idea, put forward by the late physicist and science fiction writer Robert L. Forward, was a craft pulled by a solar sail. Solar sails are a promising technology, but of course they rely on the Sun for their energy, so their use for interstellar travel isn't immediately obvious. To overcome this limitation, Dr. Forward suggested concentrating solar power into an incredibly powerful, tightly focused laser beam that could push the spacecraft all the way to a nearby star. Using a very clever design it would even be possible to slow the craft and bring it back. This would be done by firing the laser across interstellar space and bouncing it off a mirror on the spaceship to decelerate it and then push it back home.

This solar sail propulsion system could slowly accelerate a spacecraft up to 30 per cent of the speed of light, meaning it could reach a nearby star within a human lifetime. The problem with this system is the extreme scale on which it would need to be built. For a human mission, a large spacecraft would be required, and the lasers driving such a spaceship would use more electrical power than is currently generated on Earth. The sail for the spacecraft, and the lens that would be necessary to focus the laser on the craft across distances of light-years, would each have to be roughly the size of Alberta. Even a much smaller robotic probe would be an engineering feat of spectacular scale, with a sail 100 kilometres across.

And this is one of the most promising and best-developed ideas.

So propulsion is a massive problem, and it's not the only one. Time is a big issue, even if we do find some new propulsion system that defeats all these difficulties. Let's grant the possibility that some new technology emerges that allows us to accelerate spacecraft without using incredibly huge amounts of energy. Travel is still going to be a very slow process. A ship that can accelerate at the equivalent of one Earth gravity, so that its passengers feel as if they are no heavier than they are on Earth, would take a full year to get to light speed. And travelling at light speed is impossible. Unless Einstein was wrong, nothing travels faster than light, or even as fast as light or anywhere near the speed of light. So, capping hypothetical speed at a more possible number of half the speed of light, it would still take more than eight years to get to Proxima Centauri.

Unfortunately, as was pointed out earlier, there's no reason to go to Proxima Centauri. It's a tiny star, the companion of the more interesting Alpha Centauri, which is more similar to our sun. If we want to go anywhere, it's to a solar system with a star similar to ours, ideally with a planet similar to ours. However there are relatively few stars similar to our sun in our part of the galaxy. Within 20 light-years (say forty years of travel) there are only sixteen stars similar to ours that might have Earth-like planets. We haven't discovered any yet, as we don't have telescopes capable of spotting small, Earth-sized planets orbiting around other stars. What we have already discovered, though, is that some of the most Sun-like stars have pretty hostile stellar environments. Take Tau Ceti – about 12 light-years away – for example. It was once thought to be a promising candidate for having a solar system similar to ours, but recent work by British astronomers has found that it has about ten times as many comets and aster-oids as our solar system. Any planets would be subject to massive killer impacts at about ten times the rate we have. Since life on Earth has nearly been wiped out many times by such impacts, that doesn't bode well for settling on one of Tau Ceti's potential planets.

So, we may have to go a lot farther than the nearby stars to find somewhere worth going if our plan is exploration and colo-nization. Since we can't reasonably reach even the nearby stars, the prospects for our future among more distant stars look grim.

There is, of course, some chance that we'll discover a way to travel through space that avoids journeying through the vast gulfs of interstellar space. This again is a staple of science fiction. Warp drives, hyperspace, and wormholes all are supposed to provide ways to travel faster than light by warping or folding space-time, or creating shortcuts through it. We have no reason to believe, beyond a kind of educated wishful thinking, that trav-elling by wormhole or space-warp will ever be possible. At the moment, it might as well be magic.

Of course, two hundred years ago a machine that could fly would have been considered magical. So would have been a machine that could do mathematics (a computer), and a machine that could transmit speech (a telephone), and a horseless carriage, and a human being on the Moon. According to legendary science fiction author Arthur C. Clarke, any sufficiently advanced technol-ogy is indistinguishable from magic. So perhaps it's just a matter of waiting for the appearance of that new space-flight wizardry.

Bring on the magic.

WHAT MAGIC MIGHT TAKE US

TO THE STARS?

A WARNING: the following segment contains both math and highly speculative physics. Reader discretion is advised.

We live, as far as we know, in a universe that is governed by Einstein's Special Theory of Relativity, which rules out spaceships travelling through space faster than the speed of light. So, it's impossible that we'll be able to travel across the vast gulfs that separate the stars on time scales that are even vaguely close to human lives. The only way we're going to get to the stars is somehow to skip travelling across space. We're going to need to bypass or warp space so that the distances we traverse are, effectively, smaller. The problem is that we haven't the faintest idea how to do this. All we have right now is the vague hope of future discoveries, and we have that hope only because so far we haven't learned enough to be able to say for sure that these ideas are impossible.

To begin with, you might wonder why it is that we can't travel faster than light. The reason has to do with Einstein's most famous equation, $E=mc^2$, which shows that energy and mass are essentially the same thing (one can be converted to the other). That version of the equation is actually incomplete, or at least applies only for masses at rest. Masses in motion have more energy, and how much they have is described by a more complete version of the equation:

$$E = \frac{mc^2}{\sqrt{1 - \frac{v^2}{c^2}}}$$

For those of you who are mathematically inclined, here's what this means (those not can just skip down a few sentences and trust us). The v in the equation is the velocity of the moving mass. As that velocity increases – as a spaceship goes faster – the E in the equation – the amount of energy required to achieve that velocity – gets bigger. As v approaches the speed of light, E approaches infinity. To accelerate to the speed of light, then, requires an infinite amount of energy. Actually, you can rearrange the equation another way to see a different implication. As you go faster, the apparent mass of the spaceship you're travelling in approaches infinity (remember, the equation describes how mass and energy are the same thing). So as you go faster, you're attempting to accelerate an increasingly massive object – and good old Newtonian physics will tell you how hard that's likely to be. This also explains why even getting close to the speed of light is effectively impossible.

Well, let's leave the math behind. Relativity outlaws accelerating up to the speed of light, but it doesn't, strictly speaking, outlaw faster-than-light travel. It just can't be done by acceleration.

So how do you do it? Well, one option is to manipulate space itself. This, oddly enough, is not just consistent with relativity, but is actually an important implication of it. Mass distorts space-time, and the way we perceive it is gravity. When we jump in the air and fall to Earth, we're not really being pulled toward the planet. We're actually sliding down a bit of curved space-time toward the huge mass of the Earth. The favourite visual analogy for this is putting a bowling ball on a rubber sheet. The rubber sheet stretches downward under the weight of the ball, creating a curved bowl with the bowling ball at the centre. If you roll a marble across the sheet, it will be "attracted" by the bowling ball and roll down the slope toward it. This is a two- (well, actually three-) dimensional analogy for a planet (the bowling ball) in space-time (the rubber sheet).

Right then, so it's clearly possible to "warp" space-time with mass. How would we do it to move through space? Two possibilities have been suggested. The first involves creating a wormhole – a kind of shortcut between two points in space. The second is to create a warp drive that will cause space to distort in a useful way around your spacecraft.

The wormhole concept exists because the equations of general relativity that describe space-time allow what's called "multiply connected space." What this means is that there can be more than one straight line between two points in space. (Yes, yes, we know. Take a second and a deep breath – it's not you that's the problem, it's the Universe.) To imagine how this could work, take a sheet of paper; it represents space. Draw two points on one side of the paper. The distance between those two points is how far you have to travel in normal space to get from one to the other.

Now push the top and bottom edges of the paper together so it folds up in the middle. The two points will now be closer together. If you poke a pencil through the paper to join the two points, you've created a shortcut between them – a second straight line connecting the two points. What a wormhole might do is manipulate space-time to create that shortcut.

Natural wormholes may exist, though we've never seen one. They are predicted to exist, perhaps, at the heart of black holes, possibly leading to other universes. Whether we can create and manipulate wormholes is entirely speculative, but it has been considered, at least theoretically. It might be possible to create a wormhole using vast amounts of mass and energy and exotic stuff like neutron star matter (a super-dense concentration of atomic nuclei). It might be possible to hold one open with cosmic strings or with something called negative matter – if these things actually exist. Interestingly, these wormholes might allow travel through time as well as through space, which will create all sorts of horrible paradoxes. Do you have an urge to meet yourself at a younger age?

The other proposal for cheating space-time is the "warp drive," which relies on stretching space-time in front of and compressing it behind you. Let's use another analogy. Cut a circular rubber band and, holding each end, stretch it out. Your left hand can be your point of departure, and your right hand can be your destination. Now, find a lovely assistant and ask him to pinch the rubber band near your left hand. His fingers represent your warp-drive–equipped spaceship. Now pull your left hand away from the spacecraft, stretching that part of the rubber band, and allow your right hand to come closer to the spacecraft, compressing that part of the rubber band. Your assistant's hand, the spacecraft, hasn't moved. Yet the departure point is farther away, and the destination is closer. Space (the rubber band) was stretched and compressed – warped, in fact – but no distance was traversed because your assistant's hand didn't move.

The best theoretical model for how a spacecraft might go about manipulating space-time like this involves surrounding it in a bubble of negative energy. We've never seen or detected negative energy, let alone harnessed it, but its existence is possible in the mathematical equations governing all this.

You might have the impression at this point that all this is a bit of physics fantasy – mathematical hand-waving by people who've watched too much *Star Trek*. It's not an entirely unfair impression. The possibility of manipulating these phenomena, if they exist, is a dream at this point. It will require the Universe to play along with some pretty fantastic speculation in some dramatic ways. This, however, isn't quite as outrageous a proposition as it sounds. Dark matter and dark energy were previously unobserved phenomena just a decade ago. Now there's some good – albeit indirect – observational evidence that these things do exist and have a big effect on the Universe. Before these observations, dark energy and dark matter existed only as possibilities in Einstein's math. Similarly, our recognition and understanding of nuclear forces is only a hundred years old or so. Before that we honestly didn't understand how, for example, the Sun could burn for as long as our planet had existed. Science has predicted some pretty exotic and surprising phenomena in the past and been proven right.

Give the astrophysicists another hundred years, and who knows where they'll take us?

How will we find a new earth?

If you were to go by science fiction movies, every star out there has at least one, and sometimes several, planets circling around it. Every planet is remarkably Earth-like. Miraculously, the full diversity of alien biospheres is fully represented by landscapes within a two-hour drive of Los Angeles. This phenomenon, needless to say, is not the result of careful research by those sticklers for accuracy employed in the locations departments of major Hollywood studios.

Unfortunately so far we've discovered no Earth-like planets in the Universe, let alone southern-California-like planets. We haven't even discovered any Earth-sized planets yet outside of our solar system. But it's only been in the last decade or so that we've been able to detect planets outside of our solar system at all. And "detect" is the right word, as we can't see them yet – we can only see subtle signs that indicate that they're there. The problem is partly that the planets are small and dim, very far away, and next to something large and bright. Seeing a planet around a star is like trying to see a mosquito hovering next to a baseball stadium floodlight from a thousand kilometres away. In the future we may be able to design telescopes to tease out the faint light of these planets in other solar systems – so-called extra-solar planets – but in the meantime, astronomers use clever techniques to detect planets indirectly and extract as much information about them as can be gleaned from the limited data we can collect. This at least provides some information until our telescopes catch up to our ambitions.

Several different techniques have been used to discover planets, but only one of them is of any real help in finding Earth-like planets. The first looks for planets orbiting pulsars – a kind of neutron star. Neutron stars are the super-dense husk of a star that's exploded or, as astronomers say, that's gone supernova. Part of the star blows off into space, and part collapses into a 20 kilometre diameter ball of neutrons with the gravity of a normal star like our sun. These tiny balls spin at tremendous speeds, and some of them – the pulsars – send out exquisitely timed regular pulses of radio waves, which astronomers can catch using radio-telescopes. In 1992, three planets were detected around a pulsar about a thousand light-years away by the way their orbits disturbed the regular timing of the pulsar's bursts of radio waves. This technique has been used to find other planets, but there isn't really much to be said of the planets themselves. While we can deduce their size (which is not too far off that of the Earth), they can't be seen easily and they aren't likely to be very Earth-like any more. The supernova that created the neutron star they orbit would have ensured that they weren't habitable.

A second technique is called gravitational micro-lensing. In certain, very specific circumstances astronomers can see light from distant stars being bent by the gravity of a planet circling a nearer star. For this to happen the stars really do have to be in alignment – to borrow a phrase from astrology. For example, a planet circling a star 17,000 light-years away was detected in 2004 because of the way it bent the light coming from a star 24,000 light-years away that was lined up directly behind it. Unfortunately, these alignments are rare and they only work with quite distant stars. As a result, this technique isn't going to tell us much about Earth-like planets either.

The most successful planet-finding technique is the one that's been telling us the most about the likelihood of finding an Earth-like planet. The technique is called the radial velocity or Doppler method. Using this method, astronomers have discovered more than two hundred planets around nearby stars that are mostly similar to our sun. We now know a lot more about how common solar systems are, and what they look like. Unfortunately, even this technique can't find an Earth-like planet, and it's not even very good at finding the kinds of solar systems that Earth-like planets will occur in. Nevertheless it's pointing at where to look when the next generation of telescopes is built, and they should be the ones that allow us to find other Earths – if they're out there.

The radial velocity technique was originally developed by Canadian astronomers Gordon Walker and Bruce Campbell, and the idea is relatively simple. Planets orbit around stars because of the gravitational pull of the stars. The gravitational pull of the planet, however, also moves the star slightly. As the planet rotates around the star, from the perspective of an observer on Earth, the star is pulled a little closer when the planet is on our side of the star, and a little farther away when the planet is on the far side. The trick, then, is to detect the wobble. This is the hard part. Planets can't move stars very much, so there's not a lot of wobble. They also do it very slowly – once back and forth over the course of a planetary year – so the wobble takes a long time to happen unless the planet is close and its year short.

Seeing a star wobble, of course, isn't as simple as watching it twinkle in the sky. What the astronomers are actually looking for is the slight change in the light coming from the star. As the star moves toward us, the light is compressed a little, which decreases its wavelength. As it is pulled away from us, it's extended a little, increasing the wavelength. So what the astronomers are looking for is a star whose light gets slightly bluer (higher wavelength) and then slightly redder (lower wavelength). The bigger and faster the wobble, the more the starlight changes. So this technique works best at discovering large planets in fast orbits close to the star, because the star is being tugged significantly, and because the fast orbit makes the shift relatively quick.

This method has resulted in the discovery of many systems that have one or even a couple of very large planets orbiting very close to their star. In our solar system, this would be like having planets the size of Jupiter and Saturn orbiting as close to the Sun as Mercury or even closer. This doesn't necessarily suggest that most solar systems are like this, but is simply a bias in the method itself. Unfortunately, a smallish planet (the size of Earth) that's in orbit at the same distance from its star (as Earth is from ours) perturbs the orbit of its star very little. As a result, this technique can't find planets like Earth. It even has difficulty finding a large planet like Jupiter if it is as far away from its star as Jupiter is from the Sun, because that creates a small, slow wobble that's difficult to detect. What we're finding, then, is what we hope are slightly odd solar systems. We hope they're odd because from what astronomers know about planet formation, these systems might well be dangerous places to hide an Earth-like planet. It's thought that the large planets we're now finding close to their stars had to have formed farther away, and then were moved inward, perhaps by gravitational interactions with other big planets. In moving in toward their stars they would have smushed Earth-like planets like grapes on a highway. But don't despair: one super-computer model has shown that it is possible that the migration of a large planet inward could be associated with the creation of a smaller, more distant, possibly Earth-like planet.

While the Doppler technique can't find Earth-like planets, it can point us to likely systems where we can use another indirect technique to search for evidence. On rare occasions we discover a planetary system in which the giant planet isn't just causing the star to wobble, it's actually partially eclipsing it during its orbit. The planet passes directly between us on Earth and its star, in what's called a "transit." These transiting planets block a fraction of the light of the star, and astronomers can detect the difference. The planet's orbit should be very steady, so the timing of these transits and the variation in the light will be very precise. Astronomers use this when looking for an Earth-like planet, because if there is a small variation in this timing, what it may be showing is the tiny tug of an Earth-sized planet on the giant's orbit. This is the technique that Canada's MOST satellite is using in its search of the skies.

In 2006, a European Space Agency satellite called COROT was launched that will also use the transit technique to search specifically for rocky planets just a bit bigger than Earth. NASA will be launching the Kepler mission in 2008 that will be far more sensitive and might even detect the transits of Earth-sized planets that cause variations of just one part in ten thousand in the light from their stars.

The one lesson we've learned from all this work so far is that solar systems are common, and we should think of this as good news. Somewhere between 5 and 10 per cent of Sun-like stars surveyed so far have planetary systems that we can detect. It's quite possible that far more have solar systems arranged like ours (small planets in close, large ones farther away), which we can't yet detect, but which are more likely to host a really Earth-like planet. The real breakthrough will happen when new super-powerful space-based telescopes, like the ones both NASA and the European Space Agency are hoping to build, come on-line, probably by the end of the next decade. These telescopes will be designed to screen out the light from stars in order to see directly (not just indirectly detect) the small, dim planets around them. Not only will these telescopes be able to find planets, they'll be sensitive enough to determine whether they have atmospheres and even of what they are made. In short, these telescopes should be able to answer in considerable detail whether there are Earth-like planets around nearby stars.

One issue we need to touch on is just what we mean by Earth-like. Earth, prior to the development of life, was a very different planet than it is today. Even the chemistry of its atmosphere was different because, before the development of photosynthetic bacteria, our planet had little oxygen in its air. We might, then, find planets like Earth, but without life and there-fore unchanged from the barren, rocky ball that Earth once was. They might be too close to their stars, and so too warm, like Venus, which suffered from runaway greenhouse warming because there were no photosynthetic organisms to take up carbon dioxide in the atmosphere. They might be too far, and now be ice-balls. These planets would be interesting, but not hospitable places to visit or live without a lot of fairly intense work of terraforming.

Then again, we might well find a planet very like Earth, with oxygen in the atmosphere, for example. If we find something like that then it's possible – even likely – that we're looking at a planet that's already got life on it making that oxygen. That would certainly be an interesting place to explore.

Who knows, in places, it might even look like southern California.

WHERE IS THE UNIVERSE'S
BEST REAL ESTATE?

AS FAR AS WE KNOW, the only habitable planet in the Universe is Earth. Astronomers, however, are taking a considerable interest in looking for other potentially habitable planets. This wouldn't be a summer place for humans, as the difficulties of space travel would make the commute impractical. Without technological breakthroughs we have trouble even imagining, we can't think of colonizing any Earth-like planets we might find. For now, astronomers are mostly interested in finding one to learn more about our own – how it formed, transformed, and developed life. Life, of course, is what we really want to find. If we could discover a world with even simple life it might give us some idea about how common life is in the rest of the Universe. A long-shot possibility is that the worlds we discover might be places where other intelligent species have evolved. In that case we'd just hope that they find us at least mildly interesting, rather than tasty.

It's only been slightly more than a decade since the technology evolved to allow us to see whether other stars had planets. Since then, we've discovered more than two hundred of these extra-solar planets. Sadly, however, so far none of them look particularly homey. Almost all of them are giant balls of gas, like Jupiter or even larger, orbiting so close to their parent stars that you could melt metal in their atmospheres. However, it's likely that in the next few years we'll discover some planets with potential. In 2007 we came close with the discovery of the first extra-solar planet that looks like it could host life. It's called Gliese 581c (a name only a real estate agent could love), and while it's larger than the Earth, and circles a dim red dwarf star, even with those disadvantages it's small enough and clement enough that life could exist there. The discovery of Gliese 581c, however, has only made the search more exciting. Astronomers have been developing their ideas about just what kind of conditions need to exist for a habitable planet to form and thrive. They've come up with a way of narrowing down the places to look for the right kind of planets – a kind of checklist for the Universe's real estate market.

Let's work through the basic requirements. Like any real estate market, there are three vital elements to finding a dream home in space: location, location, and location. It's all about the neighbourhood.

First you have to be in the right kind of galaxy. There is considerable variety in galaxies, but the galaxy must be of the right age and, to some extent, in the right part of the Universe. The issue here is that habitability depends on having the right sort of material to form the right kind of stars and planets. Most of the Universe has lots of hydrogen and helium, out of which stars can be made. For a rich planetary environment you need other materials. You need a solid foundation, made of elements like iron, magnesium, and silicon. Insulation is important, of course, so we need oxygen and nitrogen. A nice garden will need lots of carbon, and it would need watering, so liquid water would be a big advantage. Finally, the roof would have to be in good shape, to shield you from the bad weather. In this case it's space weather in the form of radiation you'd be most concerned about, and your "roof" would be a magnetic field like the one Earth has, which is generated by its rotating core of iron and nickel.

The problem is that these essential elements don't exist in the right amounts in all galaxies. They're forged by fusion during the death throes of a massive star – in a supernova explosion, in fact. A lot of supernovae have to go off for a long time to enrich the galactic dust and gas sufficiently to form habitable solar systems. This lets out all young galaxies – they haven't been around long enough to forge the key elements. It also lets out some of the less common types of galaxies – elliptical and dwarf galaxies for two – because they simply don't have the right number of the right kind of stars. That leaves the spiral galaxies, like our Milky Way. Only the largest and brightest of these galaxies have enough supernovae to make adequate amounts of the elements to form a star and solar system that could support life as we know it. That means that a large number of galaxies won't, in fact, have the right conditions for habitability. We may be down to less than 2 per cent of the galaxies in the Universe.

The chances that we're going to visit other galaxies are remote. But we already know the galaxy we're in is habitable – so let's just look around here for the best spots. Unfortunately, it turns out that for several reasons, only a small part of our galaxy is going to be suitable. There's a lot of lousy real estate in the Milky Way. The first issue is, once again, materials. Because of the way that gravity and magnetic fields shuffle material around, only portions of the Milky

Way have the right kind to make the right kind of stars and planets.

Then there's the issue that plagues so many of us. We find a good house, on what looks like a good street, and then we meet the neighbours. You don't want to live in a crowded part of the galaxy. Downtown isn't where it's at. The dense, central part of the Milky Way has simply too much going on for a simple Earth-like planet to exist and life to thrive. The black hole that resides at the centre of our galaxy gives off bursts of dangerous radiation from time to time – think of them as belches as the black hole swallows matter – and these are best avoided. There are also lots of stars in the centre of the galaxy, some of them in strange, wandering orbits. These have a tendency to come visiting, and if they get a little too close they could disrupt a planet's orbit catastrophically, or toss a few killer comets or moons toward us for a planet-shattering collision. With all those neighbours you have more chance of a nearby supernova going off and incinerating you or bathing you in killing radiation. So it's much better to move into the quiet suburbs.

The suburbs of the Milky Way are in what's known as the "disk" of the galaxy. Imagine the galaxy as a fried egg floating in space. The yolk is the galactic centre – the busy downtown. The egg white is the disk of the galaxy. Above and below the fried egg and stretching out beyond the white is the galactic "halo." Think of this as the countryside. It's unfortunately not a good location either. Again there just isn't enough of the right kind of material for the right kind of star formation.

Our solar system is, in fact, in the best possible neighbourhood – toward the outer part of the galactic disk. It's prime real estate for a young planet looking to raise intelligent apes. And fortunately we got in before the market got too high.

HOW EMPTY IS SPACE?

IT'S PRETTY CLEAR that there's not much in space. It's the big empty. Apart from a few galaxies, space is remarkably empty. Astronomers don't entirely regret this, because the fact that there is largely nothing between us and the rest of the Universe is one of the reasons we can see so much of it, thus keeping all those people looking at distant galaxies interested and employed.

However, we should be careful not to confuse mostly empty with entirely empty. While there isn't much in space, there is something, and even when there isn't something, space seems to make something of itself. If you think that's confusing, wait until we try to explain it. Here we go, but let's begin with a relatively simple exercise. Take a volume of space and measure what's in it. One problem is that all volumes of space are not created equal. Space has degrees of emptiness.

Let's start with the space near us. The empty space around wherever you're sitting right now is not empty at all. It's full of air (assuming you're not reading the waterproof edition of this book while scuba-diving). Each cubic metre of air around you contains roughly 1.2 kilograms of air, which contains about 10 trillion trillion molecules. That is a very large number of molecules indeed, so it's clearly not very empty. However, as we move upward that density drops off quickly. At the top of Mount Everest we're down to around 4 trillion trillion molecules. From there it really starts to drop off. At the official border of space, 100 kilometres up, it's down to only a million trillion molecules per cubic metre. That's not enough pressure to support a wing in flight. At the orbit of the International Space Station, roughly 330 kilometres up, we're down to only 10 trillion, which is thin enough for humans to feel it as a complete vacuum. Step out into space unprotected and you'll notice the relative emptiness of space because things in your body will try to fill it. Contrary to popular myth and the occasional movie special effect, you wouldn't explode. The pressure in your lungs would rupture them unless you let the air escape into the vacuum. You'd probably stay conscious for about ten seconds or so and you might well live if you were brought back into your spacecraft within about a minute. It's not a thing you'd want to try.

The atmosphere continues to thin as you move away from Earth, until you arrive in the part of the Solar System, nearly 100,000 kilometres from Earth's surface, where the density of particles isn't dominated by our atmosphere. This empty space is dominated by the solar wind, which is the constant stream of particles blown off the Sun. The solar wind is variable, since it can blow harder or softer depending on how the weather on the Sun has been – solar flares and storms tend to whip up the wind. It's not, however, a wind you're likely to feel. The solar wind consists of mostly single atoms stripped of their electrons by the Sun.

Most of it is just single protons – hydrogen atoms less their electron – but there are some slightly heavier atoms as well. This is thin stuff indeed. We're down to about seven million particles per cubic metre. That might sound like a little or a lot, but in fact, if you take a cubic centimetre of this kind of empty space – about the size of a cube of sugar – you're only looking at seven atoms. That's compared to 10 million trillion in a sugar-cube-sized dollop of air on Earth.

The solar wind peters out considerably by the time it reaches the edge of our solar system. At Pluto, it's down to about eight thousand atoms per cubic metre, or one atom for every 125 cubic centimetres. Just before the edge of the Solar System, it's about a thousand atoms per cubic metre. Interestingly, there it gets a bit of a bump up, thanks to the fact that the whole Solar System is moving through space at a good clip. We're travelling at about 25 kilometres per second in the direction of the constellation Scorpio. As we move, the solar wind forms a Heliosphere around the Solar System (you did read the answer to Question 22, didn't you?). At the Heliopause, the Heliosphere is compressed by the interstellar medium. We don't know how much thicker it is at the Heliopause, but the *New Horizons* space probe might tell us when it arrives there sometime after 2020. The Heliopause is not a hard edge, to say the least. We're talking about two bubbles of nearly empty space defined only by the interactions of single atoms travelling in opposite directions, and occasionally getting close enough to affect each other. It's like two clouds having a head-on collision, except much, much less dense.

Now we're into the interstellar medium. Actually, just outside our own solar system, we're in a special place. Thanks to the local interstellar weather, we're in a big patch of clear skies called the Local Bubble, an area about 600 light-years across that is particularly empty for our galaxy. The little gas and dust in it is very thinly spread out, amounting to about a thousand atoms in a cubic metre of space. Astronomers aren't entirely sure why the Local Bubble exists – it's thought that it may be the effect of an old supernova. As we move through space, we're moving toward the edge of the bubble, and we'll likely encounter areas of higher density in a few tens of thousands of years. This probably will have some effect on our solar system. The Heliosphere acts as an energy shield, a solar cocoon that partly protects us from inter-stellar cosmic rays. If the Solar System should pass through denser space, with more particles and cosmic rays (astronomers are particularly concerned with a region of denser gas called the Local Fluff), it could compress the Heliosphere. We'd then be more exposed to interstellar radiation than we currently are.

Outside the Local Bubble, the average density of gas in our galaxy is about one atom per cubic centimetre. That's a lot denser than in the bubble, but it's still pretty empty space. In fact what that means is that if you travelled across the galaxy ten thousand times, you'd hit about as many atoms of gas as you would if you walked one metre on Earth. That, of course, is true only if you didn't run into a star or fall into a black hole, but chances are pretty good you wouldn't. Space, as we're in the process of pointing out, is pretty empty. However, the density of the galaxy's gas isn't uniform. It clumps within the galaxy. Some star-forming regions have more of it. Some regions (like the Local Bubble) have less. It's generally denser toward the centre of the galaxy than out in the fringes (where we are). And of course as you get closer to the edge of the galaxy, it becomes thinner and thinner, until eventually you're not in the galaxy any more. You're not even between stars any more, you're between galaxies – in intergalactic space.

Space between the galaxies is very empty indeed, but like space between the stars, there are variations. As we look out into intergalactic space, and because of the tremendous distances involved we're looking back in time, we can see many different kinds of space environments. There are more galaxies like ours, of course, where most of the matter has coalesced and formed billions of stars; there are young galaxies just coming together, with lots of gas and fewer stars; and there are clouds of dust and gas that may be galaxies in the future. These galaxies and clouds also clump together in long streams of matter, so on the largest-scale maps of the visible Universe we've constructed, it looks like an intricate three-dimensional spider web, which represents how gravity (and possibly other influences) have structured the mass of the Universe. The emptiest space exists outside the webs and filaments, where gravity has drawn away nearly every atom. In this rarefied environment, there are only ten atoms in a cubic metre of space. If every one of those atoms were an atom of oxygen, you'd need to gather every atom in a volume equal to ten times the size of Earth in order to take a single breath of air.

That's how empty space can be.

How fast are we moving
through space?

You think you're sitting still in your chair reading this book but you are in fact careening through space at a truly unimaginable speed. Were you to come to a sudden stop there would be a vast release of energy – about one-quarter of the energy of the atomic bomb dropped on Hiroshima. This calculation, however, relies on a complicated picture of how we're moving, and how our movement is being measured. As Einstein discovered, all motion is relative. In order to understand our motion we need to decide what our motion is relative to.

What we need, then, are mileposts. Speed needs a reference point, but there are no absolute reference points in the Universe. There is no universal starting point that we're all moving toward or away from, there are only the other things in the Universe, which are moving relative to us. Measuring motion depends on your frame of reference. You could insist that you are your own frame of reference and you're not moving at all. Thus everything else in the Universe is moving relative to you. This is, unfortunately, not a useful frame of reference, though there is no shortage of people who think that they are the centre of the Universe.

So let's take our rules as established and get down to the nitty-gritty. What we need to do is break this problem down a bit. We're moving in a lot of different ways. The Earth is rotating, so we've got the speed of rotation. We've also got our motion around the Sun – our orbital velocity. Add to that then the speed of the Solar System through the galaxy. Then there's the speed of the galaxy through intergalactic space, which gets a little complicated because it depends on whether you measure our motion relative to other nearby galaxies, or ones farther away, or some other frame of reference (which we're keeping up our sleeve for the moment).

The first part of the answer has to do with where you are on Earth. The Earth is rotating at quite a good speed around its axis. Since the Earth has a diameter of 40,000 kilometres and does one rotation a day, at the equator the speed of rotation is a little more than 1,660 kilometres an hour. This is pretty quick. It's supersonic military-jet-with-full-afterburners quick. Thankfully, everything else on the equator around you is rotating that fast as well, including your deck chair, the beach it's sitting on, your cocktail with the little umbrella and the gentle tropical breezes lulling you to sleep as you careen around on this giant merry-go-round.

That speed, however, is accurate only at the equator. The farther from the equator you are, the slower you go because the circle you describe as the Earth rotates is smaller. For example, at the latitude of Toronto (rumoured by some to be the centre of the Universe) the circumference of Earth – how far you'd have to go to circle the Earth at the same latitude – is about 29,500 kilometres. This lowers the speed of rotation to about 1,230 kilometres an hour. Go north or way south and you slow down further still, until at the poles your rotational speed is actually zero. You're still rotating once a day, but like a child standing in the middle of a merry-go-round, you're turning on the spot. This interesting gradient created by the rotational speed of the Earth is quite important and has significant influence on the weather. For example it causes hurricanes to rotate because the southern end of the storm is going faster than the northern end.

Your latitude isn't the only thing that affects your rotational speed on the Earth. Your altitude does as well. The higher you are, the faster you're going, no matter whether you're climbing a mountain or sitting on the top of a skyscraper, since you're effectively increasing the circumference of the circle you're on as the Earth rotates.

There are some other, non-speed related effects of the Earth's rotation. One is on your weight. At the equator, gravity is pulling you toward the planet, but the speed you're moving at gives you a certain centripetal acceleration. In other words, the rotation of the planet is trying to throw you off. Thankfully, gravity is a stronger force, so we don't lose anyone this way, but what it does mean is that you're actually lighter on the equator than anywhere else. As you travel north, your speed decreases, so does the centripetal force, and so your weight increases. The total difference between the equator and the North Pole amounts to between 3 and 5 per cent. So if you want to lose a quick couple of pounds, take a trip to some sunny resort closer to the equator. This, of course, also explains why you always find you've gained weight when you come back from vacation.

As fast as it can be, the speed of the Earth's rotation is actually not significant compared to the other kinds of speeds at which we're moving. We also circle the Sun once a year on a long orbit of roughly 940 million kilometres. This puts our speed in orbit around the Sun at an almost preposterous 107,000 kilometres an hour. This is far faster than any speed that humans have achieved in any vehicle. The fastest manned vehicles were the Apollo missions, which didn't quite make 40,000 kilometres an hour relative to the Earth on their return from the Moon. If you were to travel from Toronto to Vancouver at this speed, your flight time would be two and a half minutes. The flight assistants would barely have time to tell you how to put on your seat belt.

The next level of speed is the velocity at which the Solar System is travelling around the galaxy. The galaxy is rotating at a pretty good clip and our sun's orbit around the centre of the galaxy takes about 250 million years. That's a long time for a rotation, but the galaxy is unimaginably huge. We're about 26,000 light-years from the centre of the galaxy, and so our orbit around the galaxy has a circumference of roughly 163,363 light-years ($2\pi r$ for those calculating at home), so we're travelling about a million trillion kilometres in 250 million years. (You can see why astronomers use light-years as a measure of distance. Millions of trillions of kilometres start to be a little unwieldy, but we'll stick with them for our speed calculations.) This works out to a rotational speed around the galaxy of about 700,000 kilometres per hour. That speed would get you from Vancouver to St. John's in 37 seconds, and from Earth to the Moon in half an hour. Mars would take a bit over three days.

Calculating our speed relative to things outside our galaxy is a little complicated. The Universe, as you might have heard, is expanding and has been doing so for about 14 billion years since the Big Bang. It's easy to misunderstand what this means, though. The image that almost invariably (and quite naturally) comes to mind is that the Big Bang was a kind of explosion, with everything flung – and still flying – outward from a central point. But this is not the way it works, which leads to a lot of confusion. When astronomers look into the sky they see no centre of the Universe. In every direction they look it seems to be more or less the same. The picture they've arrived at is not of galaxies flying apart, but of the space between the galaxies expanding – or at least some of them, as some are still bound together by gravity, and so are actually flying toward each other. This concept makes relative speed a strange notion to work with at the large scale of galaxies. It also leads to some fairly odd conclusions, like the idea that, in some sense, there are galaxies moving away from us faster than the speed of light. We all know that we can't travel faster than the speed of light – nothing can. So that means there's something wrong with this picture.

In fact, this whole concept tends to result in migraine headaches for many people, so it's sensible to take a precautionary analgesic. There are some straightforward parts to this story. Think of the fate of the Universe as being determined by gravity and expansion. Gravity, of course, is the force that pulls masses together. It is strong when things are relatively nearby, but weakens over distance. If masses are close together, gravity can dominate and pull them together. Now for expansion. Expansion is growth of the space between masses that are far apart enough that gravity is weak. These two forces have determined that the history of the Universe is of local clouds of mass coming together (forming galaxies or clusters of galaxies) while the space between these concentrations of mass grows larger.

Our Milky Way galaxy is part of a group of about twenty galaxies called, unimaginatively, the Local Group. Thanks to gravity, these galaxies are all moving toward each other at a speed of about 144,000 kilometres an hour. So that's a good speed for us. Pretty much every other galaxy in the Universe, however, is moving away from us as the Universe expands. Since everything is moving away from everything else, what this means is that the farther away from us a galaxy is, the faster it is moving away from us. In fact, the farthest we can see in the Universe is the afterglow of the Big Bang called the Cosmic Microwave Background radiation, or the CMB. It is the first light in the Universe, from a time before there were any stars or galaxies, and it is everywhere around us. Astronomers have actually calculated our speed relative to the CMB at about 600 kilometres per second or 2.2 million kilometres per hour. This you could say is our current speed relative to the closest thing we have to a starting point in the Universe. By the way, it's this speed, and your weight, that gives us the kinetic energy of your stopping dead that we mentioned at the beginning of this answer. It's equivalent to a small nuclear explosion.

We're not quite done yet. Measuring Earth's speed relative to the CMB gives us this enormous figure, but what about measuring us against things moving faster than the speed of light? Well, we can look out into the sky in any direction through our best telescopes and see light from things that are a little more than 12 billion light-years from us – the oldest, earliest galaxies. So one way, we can see 12 billion years' distance. The opposite way, we can see another 12 billion years' distance. Logically, then, these galaxies are at least 24 billion years apart – a little more actually, since they've had 12 billion years to move apart since they emitted the light we're seeing. The problem is that the Universe is only about 14 billion light-years old. To get 24 billion light-years away from each other, since everything in the Universe started at the same point at the Big Bang, these galaxies must have travelled apart faster than the speed of light. By the same logic, there must then be galaxies 24 billion light-years from us (which we can't see because their light hasn't reached us yet), so we must have travelled faster than the speed of light relative to them. Except we didn't. What really happened was that space expanded in the voids separating all the galaxies, and the expansion of space is not the same thing as the velocity of things flying apart. It's not speed, it's growth. As soon as the Big Bang happened, there were already things so far apart they could never see each other because of the inadequacy of the speed of light.

So we can't quite say that we're travelling faster than the speed of light. Space has expanded faster than the speed of light, and we're in space, but that's where it stops.

Might be time for a more powerful painkiller.

WHY ARE THE STARS BLUE?

WELL, THEY AREN'T, or at least not all of them are. Most of them are red or brown, depending on how you count them. A few are yellow – like our favourite, our sun. Just a few, in fact, are blue, though the blue ones are very impressive. Stars actually come in many colours: yellow, brown, red, white, black, and colours we can't even see. There are an infinite number of stars, as far as we know, and in that infinity is a bewildering variety of sizes and shades. These all reflect the different conditions under which the stars were born, and also how they'll mature and die, so seeing the colour of the star is a useful way of understanding just where it's been and where it's going.

This is all because the colour of a star is a reflection of a host of features, including its size, its age, and the elements that it contains. Temperature is most important of all, but all of these factors are related. Size determines to a large extent how hot a star will

burn, how long it will live, and just what kind of catastrophic end
it will come to (things always end badly for stars). The larger the
star, the sooner and more violent its demise will be. In this sense,
big real stars are like some big movie stars – they live fast, they die
young, and they often go out in a blaze of glory.

So here's a brief guide to the colours of the stars. We'll
save the biggest and most dramatic stars for the end and begin
with the littlest stars.

Brown dwarf stars get no respect. They're even sometimes
called failed stars. They are also a relatively new member of the
Universe's family, at least to us, as the first one wasn't discovered
until 1995. Since then, though, many hundreds have been found.
Brown dwarfs are what can result if a growing star is deprived of
the resources it needs to thrive. A brown dwarf forms much as a
larger star does – from a cloud of interstellar gas – but its cloud is
just a little too small. As a result the dwarf never grows big
enough for its internal heat and pressure to start the nuclear
fusion reaction that drives larger stars. That size limit is a little less
than a twelfth of the mass of our sun. Brown dwarfs can be much
smaller – as small as eight times the mass of Jupiter (that's still
2,400 times the mass of the Earth). Brown dwarfs can still be hot,
at least when they're newly formed, because they have all the
energy of the collapsing cloud of gas that they formed from, and
so they glow – dimly. This heat allows them to have light, but no
nuclear fusion. With no fusion within, they start to cool as soon
as they form as they radiate what heat they have away. As they
cool they shrink and become even more dim. Brown dwarfs aren't
really brown. They are red, but the name red dwarf was already
taken (see below) and brown was the next closest colour.

Once thought to be rare, brown dwarfs may be among the
most common stars in the galaxy. Many of the ones we know

about are in orbit with a larger star – forming a binary star system, which turns out to be quite a common arrangement. Solo brown dwarfs have been found as well, but since they are so small and dim, they're very difficult to spot. Many astronomers now think that the galaxy may be full of millions of small, cool dwarfs that are drifting, nearly invisible, through the skies. All in all, there may be as many brown dwarfs – failed stars – as there are successful stars. New telescopes designed to look for their feeble glow in the infrared – a lower wavelength light than visible light, and the kind of light we perceive as radiant heat – may confirm this in the future. One interesting finding in the early 2000s was that, despite their tiny size, brown dwarfs can have solar systems. As they form, planets can form around them – possibly even Earth-like planets. For a time – before it cools – a brown dwarf could provide enough heat to a planet for it to maintain water oceans. So it's possible that a habitable planet orbiting a brown dwarf could exist briefly. It would soon freeze, of course, as the dwarf star cooled.

Stars the next size up are the red dwarfs. Red dwarfs are larger than brown dwarfs but only just, so if brown dwarfs are failed stars then red dwarfs scraped by with a D-minus. While we think brown dwarfs are common, we know red dwarfs are. Red dwarfs range in size from about 8 per cent to a little less than 50 per cent the mass of our sun. They represent up to 80 per cent of the stars in the galaxy (not counting those "failed stars," the brown dwarfs). Unfortunately, in our galactic neighbour-hood red dwarfs are too distant, and thus too dim, to be seen with the naked eye. Our night sky is full of them, but we can't see them without the aid of a telescope, though binoculars will do for the very brightest of them. Red dwarfs are the marathon runners of the galaxy. They burn slowly, taking perhaps a year to emit the energy our sun radiates in a day. As a result, they are expected to burn for a very long time – possibly up to hundreds of billions of years or even a trillion years, which is at least ten times and perhaps a hundred times longer than our sun is expected to be around. Since we think the Universe is less than 14 billion years old, there are no old red dwarf stars as yet, just youngsters. When they do burn all their nuclear fuel, the red dwarfs will collapse into white dwarf stars (we'll get to those). Like brown dwarfs, red dwarfs too have been discovered with planets around them. In 2007 the most Earth-like planet ever discovered – a little guy only five times the size of Earth – was discovered orbiting a red dwarf about 20 light-years away. It orbited its star far closer than we orbit ours and thus is thought to gather enough of the feeble light from the red dwarf to make it possible that water could exist on its surface.

When a star is larger, and therefore hotter, than a red dwarf, it enters the realm of the Sun-like stars, which vary in colour depending on their temperature and size. The smallest, which are a little smaller than our sun (there's no precise cut-off), are cooler and more orange than our sun. Our sun is a small yellow star – officially classified as a yellow dwarf. Larger and hotter stars are whiter, and the largest emit light in the blue part of the spectrum. The largest and hottest stars are the blue supergiants. Stars range in size across this colour spectrum, and there is no clear dividing line between the colours. There is a clear point of division in their fates, however, as what happens to the stars as they age and die is determined by their size.

For stars up to about ten times the mass of our sun – yellow and white stars – life is long – about 10 billion years (the more massive the star, the shorter the lifespan). As the star exhausts the hydrogen in its core, it starts burning more helium and swells

into a red giant star. It expands enormously – our sun will swell
so much that it will swallow the inner planets, probably including the Earth, in its enormous girth. The swollen red giant burns
like this for perhaps a couple of hundred million years, but then
begins to run out of its remaining nuclear fuel. It burns hotter
briefly, and in a last dying burst it puffs out its outer layers to
form a giant gas cloud called a nebula as big as the entire Solar
System, but in the remnant star fusion stops. What's left is a
burned-out husk of a star called a white dwarf.

White dwarfs don't fit our size-based classification because
they're quite small – smaller even than a brown dwarf. We'll put
them here though because they are the remains of larger stars.
What's most remarkable about white dwarfs is their density.
Younger stars are like balloons that are kept inflated by the pressure created by the nuclear fusion in their cores. When a white
star has exhausted its fuel, though, it can't sustain that pressure
and it cools and collapses in on itself. Gravity – no longer
fighting the pressure maintained by nuclear fusion – asserts
itself, and the star's mass squeezes in on itself as tightly as it can,
packing the star's atoms together. The resulting white dwarf star
is quite small – typically about the size of Earth – but containing
the mass of the original star so intensely compressed that its
density is 200,000 times that of rock on Earth. Each teaspoon of
white dwarf star material has the mass of a large truck.

Curiously, the more massive a white dwarf is, the smaller it
is. Large mass means more powerful gravity, and gravity draws
the atoms together even more closely in a large white dwarf than
in a small one. A white dwarf can be very hot, but it's stored heat
from its life as an active star, and it is radiated away over time. In
perhaps 10 billion years white dwarfs become cool and dark,
transitioning into black dwarf stars – we think. We're not sure

because the Universe is not old enough for many black dwarfs to have appeared, since it takes the full lifespan of a star like ours (up to 10 billion years) plus life as a white dwarf (another 10 billion) for a black dwarf to come into its prime. Certainly none have yet been discovered. Black dwarfs may only really exist in predictions about what happens to old white dwarfs.

There is another, more cataclysmic end that white dwarfs can face. They have been observed as companions to other stars in binary systems in which two stars orbit each other. In some circumstances this will lead to the white dwarf sucking material off its companion star and thus increasing its own mass. If this material is hydrogen and helium, it can accumulate on the white dwarf and heat up until it ignites, creating a thermonuclear explosion called a nova. This can happen repeatedly as material accumulates. If the white dwarf gathers enough matter, including elements heavier than hydrogen and helium, when it gets to be about 1.4 times the mass of our sun, it becomes hot and dense enough that nuclear fusion can start burning not just the hydrogen and helium, but the heavier elements in the star – carbon and oxygen. This produces vast amounts of energy and starts a runaway process in which the elements are burned into progressively heavier elements that also ignite, releasing yet more energy. The result is an almost unimaginably violent thermonuclear explosion – a supernova, in short – that completely destroys the star. This kind of supernova turns out to be a very valuable tool for astronomers, but we'll hear more of that in the section on the expanding Universe.

There's a different fate in store for the heavyweights of the Universe, the blue stars, which are anywhere from ten to one hundred times the mass of our sun. They're sometimes called blue giants, the biggest being blue supergiants. These stars sprint through their existence, burning hot and fast. In just millions of years, these stars exhaust their hydrogen. The smallest then become large red supergiants, similar to red giants that form from Sun-like yellow stars, but much larger. These stars are large and hot enough internally to burn not just hydrogen and helium, but carbon, oxygen, and even silicon. Ultimately the fusion in these stars forms iron, which won't easily fuse into heavier elements. When enough iron has formed, these massive stars collapse on themselves with tremendous violence. The very atoms of the core of the stars become so compressed that protons and electrons combine into neutrons. The neutrons simply can't be compressed any further, and so there is a bounce-back, and all the outer material of the star is blown away in another kind of supernova explosion. This material is shot out to seed the galaxy, and the remnant – the solid ball of compressed neutrons – is a neutron star even denser than a white dwarf. It's about 40 per cent more massive than our sun, but is only a few kilometres in diameter. These neutron stars spin at a fantastic rate and emit regular pulses of energy from their magnetic poles. If these bursts happen to be directed at the Earth, we see the neutron star as pulsing beacons, and so we call them pulsars.

Larger blue stars can skip the red giant phase. When they exhaust their fuel they too explode in supernovae, but their remaining cores collapse into something even more exotic than a neutron star. When the huge mass of these stars is compressed by gravity, even neutrons are crushed and the core of the star collapses into a singularity – a single point that has all the mass of the entire star. This is a black hole. The largest blue stars skip even the supernovae stage, as even their outer shells are incorporated into the collapse that forms the black hole.

The bigger the stars, the harder they fall.

Are we really made of stars?

The idea that we're made of stars has been the subject of several pop songs, by Moby and Joni Mitchell to name two, because it is a rather romantic notion to think that we are, in fact, stardust. It's romantic, but it's also entirely true. The only reason life exists is because earlier stars have lived and died and seeded the Universe with nearly the entire periodic table of elements. Most of these elements have found a home in the biology, not to mention in the geology, of our living world. Carbon and oxygen are the building blocks of life as we know it. We're 65 per cent oxygen and 18 per cent carbon. We're also 3.3 per cent nitrogen, and 1.5 per cent calcium and contain biologically necessary smaller amounts of other elements like phosphorus, potassium, sodium, chlorine, iron, zinc, and many more. Earth itself is largely composed of silicon, iron, and aluminum, which provide the structure of the planet. In Earth's interior, radioactive elements like uranium help heat the planet, fuelling the reshaping of continents that recycles the surface of the planet. All this, of course, is necessary for our survival and well-being. None of these elements existed at the beginning of the Universe.

So where did all these elements come from? Well, from the stars, of course.

Immediately after the Big Bang, the Universe was a simple place. The only elements present were hydrogen and helium, with tiny amounts of the very light elements lithium, boron, and beryllium added for flavour. These existed in vast diffuse clouds in the rapidly expanding early Universe. The clouds, however, soon began to coalesce and clump under the influence of gravity, and just 400 million years after the Big Bang the first stars began to form from these light gases. These were huge things, perhaps a thousand times more massive than our sun. When gravity made the clouds sufficiently dense, nuclear fusion ignited at their cores and they began to burn fiercely and fast. Astronomers have seen the traces of the first light of these stars in the form of faint tracks of radiation from the early Universe. However what was most important about these stars, from our perspective, is that they soon died and in doing so created new elements never before seen in the Universe.

The nuclear fusion happening in the heart of these early stars provided their energy. It also created heavier atoms from lighter ones. The lightest elements are easier to fuse – hydrogen, which fuses into helium, is easiest of all. To go up the chain and forge heavier elements requires more heat and pressure, the kind available only in a star larger than our sun, and size was not a problem for these early stars. They fused their hydrogen into helium, and when they began to run short of hydrogen in their core, where fusion happens, they began burning helium into carbon, nitrogen, and oxygen. These elements were concentrated at the heart of the star, but circulated outward and were taken up by the solar wind and scattered into the interstellar medium. In stars larger than eight times the mass of our sun, even heavier elements were created by the stellar furnace: neon, magnesium, silicon, sulphur, argon, calcium, titanium, and last of all iron. Iron was where the process stopped. Iron is to stars what kryptonite is to Superman and what garlic is to vampires. It can stop them dead and ultimately destroy them.

Once the nuclear forge has created iron in the core of a star, it ceases to burn, because, unlike the lighter elements, the fusion of iron doesn't produce energy – it requires it. A star generates a solid iron core from lighter elements, and then it cools and collapses in on itself, compressing the iron core into a mass of protons and neutrons. At a certain point, this compression can't continue, and the material bounces back, creating a shock wave of unimaginable power. This is the trigger of the supernova. The shock wave spreads outward, speeding up as it passes into less dense material and heating it and compressing it to temperatures far beyond those that exist in the heart of the star. This is the energy that can forge the heavier elements beyond iron, such as gold, silver, and lead right up to the heaviest unstable radioactive elements like uranium. Even as these elements are forged, of course, they are blown out into the Universe by the supernova explosion.

The shock wave from the supernova hasn't done its work yet, though. While it destroys the star it was born in, it helps create new ones as it propagates outward through space. It carries atoms from its own star, but also concentrates and shapes dust and gas from other exploding stars into rich mixtures of material that become stellar nurseries. In these nurseries, atoms and gas combine into dust, and chemical reactions between these newly forged elements produce more complex molecules.

Molecular gases and water and methane form, as well as minerals that clump together to form asteroids and planets. From these clouds are born stars like our sun, surrounded by disks of enriched dust from which planets form. So the material that comprises our solar system came directly from the death by supernovae of earlier stars.

The processes that are going on in a supernova have been studied in some of the most powerful particle accelerators on Earth. These are able to produce, for fractions of a second and in tiny amounts of material, the same conditions that produce megatons of material in a supernova. By this method, physicists have studied some of the exotic and short-lived elements that previously only ever existed for brief times in supernovae explosions. So while we are all made of stars, we are only now, and only on tiny scales, making stardust ourselves to better understand just how the stars do it.

Is there intelligent life in the Universe?

We don't know. You were expecting a different answer? Well, here's a longer one:

Humans have always been consumed by the notion that there is a different kind of life in the heavens. Before we understood what the heavens were, we believed gods and spirits lived in the skies. Now that we understand – well, partially understand at least – just what space and the Universe are, we don't picture supernatural beings cavorting on clouds. Instead we speculate about other kinds of natural life, which is to say intelligent alien life. While our fantasies about alien life reflect more about ourselves than the reality of life elsewhere in the Universe, the modern concept is not analogous to the ancient gods. Science and astronomy have been put into service to support at least the speculation that there is intelligent life elsewhere in the Universe. We know enough about the Universe that we can begin to calculate just how likely the conditions are that gave rise to the only highly intelligent species we know about. By the way, just to be clear to the skeptics, that's us – not the dolphins.

This is not to say that there's any real precision to these calculations. They are for the most part only informed guesses because there are large holes in our understanding of how life arises, let alone intelligent life. If we set the precondition for life to be an Earth-like planet circling a Sun-like star, then there is a great potential for life to have developed elsewhere. Sun-like stars are not rare – about 10 per cent of the stars in our galaxy are similar to our sun. Earth-like planets are probably more uncommon, but we don't really know. We've not yet detected one around any other Sun-like star, as we don't yet have the technical capacity to do so. In the last decade or so we've discovered close to two hundred planets around nearby stars, but they're all giants – ranging in size from at least five times the size of our planet to several times larger than Jupiter. We can detect these planets with our current telescopes, but our instruments aren't sensitive to anything much smaller. NASA hopes to launch a new space telescope called the Terrestrial Planet Finder by around 2020, which should be able to detect Earth-like planets around reasonably nearby stars. Nevertheless, the fact that we have seen so many giant planets around nearby stars suggests that solar systems are relatively common, and so Earth-like planets might be as well. An Earth-like planet, of course, will have to have the right kind of minerals and the right amount of water and organic chemicals to give rise to life, but these things seem to be common enough in the Universe.

The next big question, then, is just how the chemistry of a lifeless Earth-like planet turns into the biochemistry of a planet teeming with life. This is an important unanswered question at the moment, since we don't really know how this happened on Earth. Somehow – possibly as long as 3.5 billion years ago – chemistry became biology on our planet and the first micro-organisms appeared. Biochemists have speculated about the special conditions that led to the first self-assembling simple chemicals and suggested that they were something like simple proteins, or RNA – the message carrier for DNA. Their speculation, however, is just that. We simply don't know what causes that first great leap, and so we don't know if it was a massive fluke that might never repeat itself or an inevitable process, given the right conditions.

On Earth, there were probably a couple of billion years during which life was not much more complicated than a colony of bacteria. Then about 600 million years ago, in a relatively short time, larger and more complex organisms appeared. We don't know what triggered the sudden development that led to the evolution of large and complex animals, but from then it was almost no time at all until the oceans were teeming with them and the land was colonized first by plants, then insects, and finally vertebrates – animals with backbones, who were our ancestors. Like the origins of life, and the sudden development of complexity, the colonization of land is another great leap in evolution that we can't adequately explain. The next great leap is the truly mysterious one. It was the leap to human intelligence – to the single species among the millions that have inhabited the planet that has the ability to look out into the Universe and wonder if there are any good restaurants out there.

The mystery of the origins of human intelligence or con-sciousness is similar to the mystery of the origins of life. We know it happened, and it seems to have happened only once, but we don't know whether it was a fluke or an inevitability. We don't know exactly why we developed intricate languages and the ability to make complex tools and to advance technology. We suspect that brain complexity and size (at least relative to body size) have something to do with intelligence, but what causes a large brain to become a conscious one is a mystery. The dinosaurs never developed large brains (proportional to their body size) so didn't develop a conscious intelligence before an asteroid ended their evolutionary experiment. The mammals have done quite well, though, as two lineages – the great apes and the whales and dolphins – have developed large brains relative to their body size. The dolphins actually come close to humans in brain size to body size ratio, but so far only humans have made the leap to advanced intelligence, with things like language and sophisticated tool use.

So there are a couple of big questions we need to know the answer to before we can tell if there is intelligence elsewhere in the Universe. Is it easy to create life, and is it inevitable that life becomes intelligent? There is another way, however, to determine whether there is intelligent life out there. We can go looking for it. In fact, a great many people are doing that in the project known as SETI, the Search for Extra-Terrestrial Intelligence.

Currently the headquarters for research into this is the SETI Institute, a privately funded organization that, despite its preoccupation with intelligent aliens, doesn't investigate UFO sightings or stories of extraterrestrial kidnapping. Instead it pursues a rigorous program of research centred on listening for signals from alien civilizations in space. This isn't as quixotic as it sounds. The highest-profile SETI project involves using radio-telescopes to listen for these signals. They're not looking for the equivalent of alien sitcom re-runs leaked into space. Anything like our terrestrial radio and television broadcasts wouldn't carry well across interstellar space. The frequencies we use are easily disrupted by interference from radiation. So it's unlikely that there are aliens out there trying to puzzle out the subtleties of *I Love Lucy* re-runs.

Instead, SETI researchers are listening for signals that other civilizations might have sent deliberately, using radio frequencies best capable of carrying a strong signal across space and cutting through the interference. Oddly, while we're listening for such a signal, we're not currently sending one. An alien civilization would have difficulty detecting signs of intelligent life on Earth. Our only deliberate signal meant to communicate with alien intelligence was a three-minute message sent from the gigantic Arecibo radio-telescope in Puerto Rico in 1974 aimed at a cluster of stars 21,000 light-years away. (Just so you know, the soonest an answer is expected would be the year 43,974.) Nevertheless, SETI researchers are hoping alien civilizations are more communicative than we are. SETI researchers listen for and record radio signals, which include a great deal of static, and then use powerful computers to attempt to discern patterns that might be attempts to communicate information. To date, no such pattern has been found.

This might not surprise you, but perhaps it should. In fact, there is an argument that if intelligent life does exist elsewhere in the Universe or in our galaxy, then it's possible that it is trying to communicate with us, and indeed might well be on its way for a visit. The starting point in thinking about this for about the last half-century has been something called the Fermi Paradox. Enrico Fermi was one of the great physicists of the twentieth century, a pioneer in the study of nuclear fission, and a researcher with the Manhattan Project, which built the first atomic bomb. In 1950, in the context of a casual lunch conversation, Fermi realized that while the galaxy is large, intelligent life has had a great deal of time to establish itself and spread. Any civilization would, he reasoned, reasonably quickly develop sufficiently advanced technology to travel among the stars. Humans, for example, have already got that technology, though it's in its infancy. We have, after all, sent robot spacecraft (*Pioneer* and *Voyager*) that are on their way out of the Solar System, heading toward the stars, though they won't arrive at even the nearest star system for many tens of thousands of years. Fermi reasoned that any intelligent civilization should, in a reasonable amount of time, be able to spread out and colonize the entire galaxy. By a reasonable time, he allowed a generous 10 million years. Given that our galaxy is probably upward of 10 billion years old, this is only one-thousandth the age of the galaxy. Given the amount of time available, if the chances that intelligence can arise elsewhere are better than infinitesimal, and if interstellar travel is possible, then the galaxy should be full of intelligent life. The Milky Way could be crawling with interstellar empires and civilization.

The fact that it isn't (as near as we can tell) is the Fermi Paradox. One possibility that might resolve the paradox is that we are somehow extremely lucky. Perhaps the emergence of life and subsequently of intelligence are extreme long shots, and humanity is unique in beating the odds. Alternatively, there may be something inevitable that happens to sufficiently advanced societies that lies in our future – an environmental catastrophe or a civilization-destroying conflict – that prevents societies from developing and expanding past a certain point. Maybe interstellar travel is simply too hard. Perhaps we're located in such a galactic backwater that they just haven't bothered with us.

The other option, of course, is that alien civilizations have settled the rest of the galaxy. They could be all around us, and either they're ignoring us or keeping themselves hidden from us. Maybe the conspiracy theories aren't so crazy after all.

WHAT HAPPENS WHEN GALAXIES RUMBLE?

LADIES AND GENTLEMEN, welcome, one and all, to a true heavyweight encounter. This bout is scheduled for three 250-million-year rounds, for the championship of the local group of galaxies. In the red-shift corner, weighing in at at least 600 billion solar masses – it's got a halo around it, but a black hole for a heart – it's the Milky Way! And in the blue-shift corner in the nebulous trunks, one of the few galaxies visible to the naked eye, weighing in at a conservatively estimated 400 billion solar masses, it's the Andromeda galaxy. Milky Way, Andromeda, I want a nice clean bout, now shake your spiral arms, and when the bell rings in three billion years, come out fighting.

It's true, galaxies do get it on. In fact that's an appropriate phrase, double-entendre and all, because the encounter between two galaxies could be described metaphorically either as a conflict or as some kind of more carnal interaction. However you choose to imagine it, it's important to realize that it happens, by universal standards, relatively frequently and also that it happens, by human standards, very slowly indeed.

Galaxies are generally quite distant from each other – on the order of millions of light-years apart. Most of the galaxies in the Universe are growing more distant as the Universe expands and everything becomes a little more distant from everything else. Often as not, though, galaxies are born and live in clusters or small groups that are close enough that the force of gravity is more than the force of expansion, and so they are drawn together in a complex gravitational dance. Eventually the partners can take a wrong step, and two galaxies collide. This is one of the more peculiar collisions in the Universe, as it can lead to spectacular events without anything hitting anything else. It's a non-contact bout.

There is no contact because galaxies are mostly empty space. They're not as empty as the vast intergalactic stretches in between, but still there's not much in them. Take our galactic neighbourhood as an example where crowding is about average. Within a sphere about 50 light-years in diameter there will be perhaps 2,000 stars, though only about 130 of them will be visible to the naked eye. (The others are mostly old and dim red dwarf stars and the other stars you see are farther away.) That space is so big, however, that 50 thousand billion billion stars would fit in it if packed in tight. You could fit in nearly 4 trillion things the size of our solar system (sizing it to beyond the Kuiper Belt).

What this means is that when two galaxies hit each other, it's practically inconceivable that any two stars or planets will actually collide. Far more likely is that as the galaxies interpenetrate, the stars pass by each other, missing by huge distances. However, just because the stars don't collide doesn't mean that there won't be a lot going on. Because of the power of gravity, there doesn't need to be contact for large effects to occur. Astronomers scanning the distant skies for interesting phenomena have seen many galaxies about to collide, colliding, and after collision. By taking examples from all over the Universe, they can almost generate a stop-action movie of what happens as two galaxies come together. What's more, they see it often enough that it's now assumed to be pretty normal for most galaxies in their long lifetimes to collide and to be shaped by those collisions. Galactic collisions may be one of the things that keeps the Universe's galaxies fresh and spry instead of decaying into decrepitude.

One of the things that happens in a galactic collision is that the dust has its day. While stars may miss each other, the tenuous matter among the stars has a chance to literally shine. It gets pulled about, accelerated, shocked, and jarred, and this heats and concentrates it, so it can form new stars. The pictures that our large telescopes have captured of these events are some of the most spectacular astronomical images ever taken. One kind of collision, probably the most common kind, is encounters of larger galaxies like the Milky Way with dwarf galaxies. These dwarf galaxies can collide with each other to form larger galaxies, but it's more common for them to be swallowed up by larger galaxies like the Milky Way. When they collide, big galaxies just tear small ones apart, stealing their stars away and incorporating them into their own structure. This process may well be happening now with a galaxy called the Sagittarius Dwarf Elliptical Galaxy, a small galaxy discovered in 1994 extremely near the Milky Way. These nearby dwarf galaxies can be quite difficult to spot as they are close enough that their stars can be difficult to distinguish from the stars of the Milky Way.

The Sagittarius Dwarf Elliptical Galaxy so far seems to have been a big mouthful for the Milky Way to swallow. It's orbiting around our galaxy about once in a little less than a billion years and has been doing so for a very long time, so it's made many passes. Its orbit actually cuts around the plane of the galaxy, and in doing so it comes so close to the edge of the Milky Way that some of its stars actually pass through our galaxy, perhaps as close as the same distance from the centre as our solar system is. The result of this collision – or perhaps this near miss – is that the gravity of the larger Milky Way has been ripping stars away from Sagittarius, so there is a vast stream of them trailing the circular orbit that the dwarf follows around our galaxy. This stream forms a huge, very faint ring around the galaxy. These stars will, eventually, drop into our galactic plane and become part of the Milky Way. The dwarf galaxy is fighting back as much as it can. Sagittarius's gravity has distorted the Milky Way so that its flat plane has become warped slightly.

Collisions with dwarf galaxies will cause the Milky Way to grow somewhat, but we have a much bigger collision in our future with a galactic equal – Andromeda. Andromeda is the only other large galaxy in the collection known as the Local Group – our neighbourhood in the Universe. It's a spiral galaxy like the Milky Way and might be a bit larger than our galaxy, but is probably less dense, with fewer stars so that it doesn't mass quite as much as the Milky Way. Andromeda and our galaxy are already moving toward each other at the spectacular speed of 500,000 kilometres an hour, and in about 3 billion years will collide. Neither galaxy will emerge the same.

If humans, or our evolutionary descendants, are around at the time, it will be quite a show, though they might want to watch it speeded up in stop-action as it will take hundreds of millions of years. Andromeda will become brighter and brighter in the sky, possibly becoming visible to the naked eye as a full spiral galaxy – the sort of thing we can now see only in powerful telescopes. The first visible effect will be that the spiral arms that reach out from each galaxy will tend to stretch under the new gravitational forces they are experiencing. They will lengthen out dramatically, forming huge tails that swirl around each galaxy. Some of the stars within them (including possibly our sun) will be flung out into interstellar space. Some others will be drawn into the centre of the galaxies. Computer simulations of the collision show the huge starry tails being whipped off each galaxy in a spectacular game of cosmic crack-the-whip.

Andromeda and the Milky Way will then start to pass through each other. This will completely distort their lovely spiral structures, creating a disordered mess as the stars that have been inhabiting regular orbits around their respective galactic cores are thrown about willy-nilly. Meanwhile, the dust within the galaxies will collide at a tremendous speed, forming huge shock waves. The thin clouds of dust from each galaxy will slam together like opposing hurricane winds. This will heat the dust tremendously and also concentrate it, creating new stellar nebulae – clouds of hot gas that will immediately begin giving birth to new stars. Many of these new stars will be giants, which will start to explode in massive supernovae in only a few million years after forming, so exploding stars will be flashing frequently during the extended collision. The two huge galaxies won't combine initially, though, as the speed of their collision will carry them through one another and apart again. This will be a brief separation as their mutual attraction will soon drag them back together again for a second collision, which will again mix up all the stars and give rise to new stellar nurseries and a new round of star birth and death.

After this second collision, the two galaxies will never separate again. Over time their stars will settle into a new stable order in a single giant galaxy that looks quite different from the two that birthed it. This new galaxy won't have elegant spiral arms, as it will be an elliptical galaxy – a dense cloud of disordered stars in the shape of a stretched egg. This whole process will take something on the order of a billion years.

This new elliptical galaxy (we could call it the Mandromeda Way) will likely be the last stage in the evolution of our galaxy. It will swallow a few more dwarfs, but there are no other big galaxies around us close enough to combine with. The rest of the stars in the Universe are too far away, and spreading apart too fast, for any more dramatic collisions or combinations. If the Sun survives the collision with its planetary orbits intact, and is permitted to stay within its new galaxy, this will be our home until the stars start to wink out at the end of time.

WHAT HAPPENS WHEN YOU FALL INTO A BLACK HOLE?

NOTHING GOOD, of that we can be sure. The idea of a black hole is sufficiently familiar to us that it's become a popular metaphor to describe something that swallows up anything around it. As a result, we've all got a vague understanding about certain features of a black hole: they're incredibly dense and exert a powerful gravitational pull so that anything that falls into them can't escape. They're also black – because not even light can escape them. This is all true, but black holes are even stranger than that because of what they do to space, and by extension what they do to anything that falls into them. The experience of falling into a black hole – if you're ever unlucky enough to have it happen to you – will be a very odd one. It will also, unfortunately, be painful, mercifully brief, and needless to say fatal. Nevertheless, if you insist on visiting a black hole, it's as well to have some idea of what to expect.

This presents difficulties since nobody has ever seen a black hole directly, though astronomers have detected them indirectly in enough different ways that there isn't any serious doubt that they exist. So for our guide we are relying largely on the calculations that astrophysicists have made when trying to understand just what should happen in the bizarre environment around and in a black hole.

First, a little history about the place you're visiting (though the visit is going to end up being a permanent stay). Let's start with a large star (many times the mass of our sun) at the end of its life. For millions of years it's been kept inflated, like a fiery balloon, by the energy of the fusion reaction in its heart, but now it's used up its hydrogen and helium fuel. As a result, the star begins an accelerating collapse inward upon itself, growing ever more dense. Eventually the collapse reaches a critical point. Some of the collapsing material "bounces" outward in a supernova explosion, releasing vast energy. The remainder, however, becomes ever denser. Atoms are so compressed by the collapse that electrons and protons are forced together against their natural repulsion – and they transform into neutrons. If the star is small, the process stops here, and what's left is a perfect, solid ball of super-dense neutrons – a neutron star. If the star is more massive, though, even the neutrons can't resist the collapse, and the matter is compressed into an infinitely small and dense single point called a singularity – the core of a black hole.

It's here that things start to get strange. The black hole is infinitely dense, but it doesn't have to be all that big or massive. It's actually going to be lighter than the star it came from (since much of the stuff of that star was blown away in the supernova). It exerts no more gravitational pull on the objects around it than the original star. Any planets that survived the supernova explosion now orbit a black hole instead of a large star. Things are quite different as you get close to the black hole, though.

Because the black hole is infinitely dense, the gravitational effects it produces are quite different from those of a normal star. The maximum pull of an ordinary star's gravity is at its surface, where its gravitational field is strongest, and it doesn't get any stronger closer to the centre. Because all a black hole's mass is concentrated in a single point, as you approach, its gravity just keeps getting stronger, and stronger, and stronger. At the surface of our sun, you could escape its gravity by accelerating to about 144,000 kilometres per hour. At the event horizon of a black hole, however, even the speed of light – 300,000 kilometres per *second* – wouldn't get you out.

That term – event horizon – needs a little explanation. The event horizon is what we think of when visualizing a black hole – a sphere of black in space. In fact we can't see the singularity – the heart of the black hole – at all. What we can see is the region of space around the singularity where the gravity is so strong that light can't escape it (because light, after all, can only go as fast as light). Since light can't escape it, this region looks black. We call it the event horizon because, like the horizon of our planet, we can't see beyond it. What's interesting, and what we'll explore in a moment, is that while everything inside the event horizon looks black from the outside, it wouldn't look black from the inside. As happens with the planetary horizon, travellers who disappear over the horizon can still see each other and what is around them. So if you bring company on your trip into the black hole, you won't be lonely – not that either of you would survive the experience.

All black holes are singularities, but they are not all the same. They can vary in size or, more accurately, in mass. A brand-new black hole from an exploding star will have a mass several times that of our sun, with an event horizon perhaps a few kilometres across. Then it could start to grow by swallowing stars, planets, dust, and other black holes. We now think that giant black holes with the mass of millions of stars inhabit the centres of many galaxies, including the Milky Way. Our black hole is estimated to have a mass equal to that of nearly three million suns. These huge black holes can have event horizons many millions of kilometres in diameter. Interestingly, black holes can be small as well. Theoretically, microscopic black holes can exist since it's density, not mass, that makes a black hole what it is. A tiny amount of matter forced into an infinitely small space could become a black hole. There's no way to make a microscopic black hole in the modern Universe that astronomers know of. The Big Bang, however, might have created billions of mini-black holes, which some theorists think then merged to form the black holes at the heart of galaxies like the Milky Way.

We've never been close enough to see the event horizon of a black hole, but we have seen strong evidence of them in the form of their table scraps. This is light from material that has come close to a black hole but hasn't quite fallen into it, a band around the event horizon called the accretion disk. As gas and dust particles from the accretion disk approach a black hole, they are accelerated to tremendous speeds by gravity. They smash into other particles, which are also travelling at high speed, and these collisions release powerful X-rays that are radiated out into the Universe. So while black holes can't be seen directly, we can see signs of them eating by the X-ray glow that surrounds them. The Chandra X-ray observatory – a huge orbiting telescope similar to the Hubble Space Telescope – has taken spectacular pictures of the X-ray glow from the black hole at

WHAT HAPPENS WHEN YOU FALL INTO A BLACK HOLE?   191

the centre of our galaxy, indicating that the black hole there is something of a sloppy eater.

Now you know the history and culture of the black hole – the parts of a guidebook most people skip while they look for the good restaurants and hot beaches. What do you have to look forward to as you visit your black hole?

Well, the trip there in your spaceship will be much like any trip to another star system. Black holes can be anywhere. If you go to one of the outlying suburban black holes you might see a neighbourhood much like the one around our solar system, though if it's very new it would be surrounded by the spectacular stellar nebula left over from the supernova that birthed it. Going to the black hole in the centre of the galaxy will be like visiting New York after a lifetime in Wyoming. Everything will be too bright and moving too fast. The centre of the galaxy is a busy place, with many stars packed close together around the giant black hole in the middle.

As you approach the event horizon, you'd start to accelerate under the influence of its gravity. You may go into orbit around it, beyond the event horizon, and you'll find yourself jostling with all the other matter lined up for a chance at a big exit. Unless you drop straight in, you'll also be part of a huge whirlpool of matter spinning around the black hole like water spiralling around a gigantic drain. If you're not careful this jostling will likely grind you up into a super-heated ionized gas – a plasma – because this is a very hot place to be – full of X-rays and gamma rays being emitted by super-high-speed collisions among the objects falling into the black hole. All this jostling might actually throw you away from the black hole, but there's just as good a chance that it'll throw you into it – allowing you to skip the line-up. Well before you reach the event horizon you'll be feeling uncomfortable because the steep gradient in the black hole's gravity is acting on you unevenly. If your feet are closer to the black hole than your head, then the gravity is going to be pulling on your feet more strongly than it's pulling on your head. What you're experiencing here are tidal forces similar to those that raise sea tides on Earth. Don't worry, these will get worse. You still have the chance to change your mind – reverse engines on your spaceship and accelerate away at a good fraction of the speed of light. Or you could even hover on the horizon briefly, thinking about your fate. However, if you cut your engines and slip over the edge, you're gone. You are permanently cut off from any interaction with the outside Universe ever again.

It won't suddenly go black as you pass over the event horizon, and you might see some interesting phenomena. The light from the rest of the Universe will appear twisted and distorted, perhaps shifted in colour by the black hole's distortion of local space-time. It's possible that the light reflected off the back of your head could be curved around by the gravity of the black hole so that you can see it in front of you. After all, it can't escape the black hole – it has to go somewhere. You'll still see the Universe outside the black hole, though it may look somewhat distorted. That light from outside is coming in with you, and, like you, it won't ever be leaving again. So what do you see if you look down toward the singularity? Unfortunately this is where our guidebook breaks down. We don't know. For that matter, we don't know exactly what a singularity is. It's just a name for an infinite amount of mass in an infinitely small space, and we don't have any physics that helps us explain how it works, and therefore what it might look like.

If you are only now having second thoughts about suicide and have some intellectual curiosity, you could fire up the engines on your spaceship. The acceleration will be crushing, but we'll imagine somehow you can survive it without being turned into jam. You still have a problem. Despite your acceleration, which should be slowing you down, you seem to be going faster toward the singularity. In fact, your acceleration is making your subjective time pass more quickly, effectively reducing the amount of time you have before you crash into the singularity. Cut your engines, and free-fall in – you'll gain a few precious fractions of a second. As you fall, the tidal forces that have been making you uncomfortable take over. Your legs will be pulled down, and you'll start to elongate. In no time you'll have been stretched out nearly infinitely down toward the singularity – "spaghettified," one physicist says. At this point you'll have rather lost interest in the rest of the proceedings. Rest in pieces.

And sadly this is the last we can say about your trip. We can confidently make predictions to a certain point, but then the black hole gets too weird for even physics to describe. The singularity is still waiting for an explanation. Some physicists think that exotic string theory might explain the interior of a black hole, but string theory itself is under attack these days. One irony, of course, is that it's only physics theory that will ever tell us what is happening inside a black hole. The observational astronomers are going to have to be silent. The nature of the beast is that anyone who goes to look will never return to tell the tale.

CAN YOU SURF ON A
GRAVITY WAVE?

ONE THING WE CAN THANK EINSTEIN FOR is making the Universe seem a lot weirder. The realization that space could be warped and bent and distorted, which was an essential feature of Einstein's theory of gravity, was a revolution in our understanding of how the Universe worked. It did Newton's theory of gravity one better by providing an even more satisfying (if less intuitive) explanation of the operations of the Universe. At the same time, though, it meant that we couldn't trust what we saw any more. We don't see space being warped and bent around us, and yet it is. In fact, it's not just warped and bent, it's actually moving. Einstein's theory posits that space is constantly rippling around us, being distorted, at a microscopic level, by waves of gravity. The next big breakthrough in our understanding is likely going to be the detection of these waves of gravity, and then using them to study just what the Universe is doing down in the gravity well.

To understand how gravity can cause waves, you first need to understand what gravity is. Isaac Newton invented the concept and provided the famous inverse square law that allows us to calculate the force of gravity. Newton, however, didn't explain what gravity actually was. He described, mathematically, how masses affected each other through an attractive force that was related to their distance. He calculated its strength, and in doing so explained the motions of the objects in the sky. What this force was or where it came from he didn't try to explain. That was part of Einstein's great contribution. But Einstein didn't explain gravity, he eliminated it as a force entirely. In Einstein's theory there is no gravitational "force" as such. What we perceive as the force of gravity is nothing more than the way space is warped by concentrations of mass.

If you've ever read anything about Einstein's gravity, you've probably encountered the water bed analogy used to describe how it works. It's a reasonable two-dimensional analogy for how the three dimensions (four if you add time, but we'll leave that out for the moment) of space work. Imagine space as the surface of a water bed. It's flat of course, and that represents empty space. Now take some heavy objects – cannonballs of different sizes would do well – and put them on the water bed. They can represent masses of celestial objects – planets, stars, whatever you like. The surface of the water bed is now distorted. Every ball creates a flared, cone-shaped depression around it. The larger the ball, the larger the depression. The way the balls warp the surface of the water bed is analogous to the way planets warp space. You can even see how gravity works in this analogy. Just roll a marble across the water bed. The marble could represent an asteroid or a spaceship. If it gets close to a "planet" it will, very likely, fall into the "gravity well" created by the balls. If you roll a marble on just the right trajectory, you can see how it can skirt the side of a depression, circling the ball briefly – like going into orbit.

That's all fun, but here we're interested in gravity waves. To see those, start moving those cannonballs around on the bed constantly. Think of them orbiting each other, or passing by each other. They'll set up oscillations – waves – in the water. The surface of the water bed will be distorted as the waves pass along it. This is pretty much the way gravity waves are supposed to work.

Gravity waves, however, are a bit different from water bed waves in that they don't just cause you to bob up and down. As they move through space, they do something quite strange. They increase and then decrease the size of the space they move in. If a gravity wave hit you, it would momentarily and suddenly cause two dimensions of you and the space you occupy to get larger and then smaller. In other words, if the gravity wave hit your front, you would get taller and wider, and then shorter and narrower as the wave passed through you. The effect would be much like what you see in fun-house mirrors except that it would actually be you distorting and not just your reflection.

The other odd thing about gravity waves is how minuscule their effect is. The difference between waves you might set up by cavorting in a vigorous manner on a water bed and those set up by moving planets in space are a matter of scale. As water-bed owners can testify, jump on a water bed carelessly and your pseudo-gravity wave might be enough to fling your partner out. Real gravity waves, on the other hand, are not so dramatic. While any movement of mass in space – waving your hand, driving your car, moving an asteroid – will cause a gravity wave, it will be so infinitesimally small that you're not going to feel it. The gravity waves caused by a really huge movement of matter – let's say a supernova going off in our galaxy close enough that we could see it in daytime – would be essentially unnoticeable. The last time this happened, we think, was in the eleventh century when Chinese astronomers recorded a "guest star" in the constellation Taurus. It was visible in the daytime for about three weeks. This would have been a massive explosion of a star far bigger than our sun, driving out much of its mass at speeds approaching the velocity of light. This tremendous disturbance would trigger a "gravity tsunami" that would cause your height and width to increase, as it passed through you and distorted the space you occupy, by one-billionth of a billionth of a metre – a fraction of the width of a proton.

Just as well. It would be alarming and probably not very comfortable if gravity waves were changing our shape by any substantial amount all the time. We'd all have to wear much more flexible clothing to accommodate the constantly changing sizes of our bodies. Oh, and we'd be torn apart – messily.

Now for the embarrassing admission: gravity waves are so small, and their effects so subtle, that astronomers have never actually detected one. They're pretty sure all this gravity wave business is happening, but they've never caught it in the act. Einstein himself – even as he developed the theory that would predict gravity waves – doubted that they would ever be detected as their effect is too tiny. In that case, how much do they matter in the operation of the Universe? The answer is that they do have a meaningful effect, but only in the most extreme circumstances. Since gravity waves represent a form of energy, as all waves do, then they are, in some sense, one way in which the Universe is winding down. They play a part in the dissipation of all energy – the slow cooling that is eventually going to leave the Universe cold, dark, and boring, though not for many, many billions of years.

We know this because the only observational evidence we have so far for the existence of gravity waves is the mysterious loss of energy from binary neutron stars. Neutron stars are super-dense accumulations of matter, just a bit too small to form into black holes. There are rare instances of two neutron stars orbiting each other. Astronomers who have measured the orbits of these binary neutron stars noticed that the stars are falling toward each other – their orbits are gradually losing energy. The only place this energy could be going is into gravity waves. This discovery won two Princeton scientists the Nobel prize for physics in 1993, so you can see how important physicists think it is.

The next Nobel prize having anything to do with gravity waves is likely going to be awarded to the first person who directly detects them. There are several groups of physicists using super-sensitive instruments in the hope that they'll soon detect a powerful gravity wave – perhaps from colliding neutron stars, or black holes or supernovae or some cosmic smash-up. The most sensitive detector in operation these days is made up of two giant L-shaped laser beams with arms four kilometres long. The lasers are precisely tuned and timed so that if a powerful gravity wave "stretches" them bells will ring and gongs will sound and someone will jump on a plane to Stockholm to pick up the prize. There's been nothing yet, though, perhaps because the detector is not large and sensitive enough. So there's hope of making a space-based detector built on the same principle, though much larger, in the future.

The reason astronomers are so interested in detecting gravity waves is not just to prove a theory right and collect prizes. Gravity waves, like light waves, can be a way of studying the Universe. Astronomers use all sorts of electromagnetic radiation – from infrared radiation to X-rays – in their observations. Different objects emit light in different ways, and so every frequency of light brings new information about things that might never have been visible before. The dream of gravity wave researchers is to be able to see the Universe "illuminated" by gravity rather than light. That would reveal the secrets of some exotic phenomena – things we can't see well in light like black holes and the echoes of the extreme gravitational disturbances created in the Big Bang. Indeed, you might say that the gravity of the situation will cause us to see the whole Universe in a new light.

What about surfing on a gravity wave? Sadly, the answer probably is we can't. Gravity waves are expected to move at the speed of light; surfing requires the surfer to ride with the wave at the speed of its propagation; and we know of no way to get a surfer up to the speed of light to catch the gravity wave. Interestingly, if you somehow managed to develop a spacecraft that could approach the speed of light (it would require an almost infinite amount of energy to accelerate it, but we'll leave that for the moment) you would actually start to perceive gravity waves. You would, in fact, start banging over them like an old pickup over a wash-boarded road.

Which might just make gravity waves the Universe's ultimate speed bump.

WHAT'S ON THE UNIVERSE'S
RADIO STATIONS?

RADIO, AS ANY *Quirks & Quarks* listener knows, is a universal medium, and it's also the medium of the Universe. The Universe is full of radio, and listening to it is one of the best ways to learn about the history, current events, and even the future of our solar system, galaxy, and all the strange and wonderful things farther afield.

Unfortunately the Universe's radio doesn't have (as near as we've heard) what we normally think of as radio stations. Nearly every object in the Universe that emits visible light (and many that don't) also emits radio waves. This radio "glow," like the glow of visible light, can tell astronomers a great deal about the nature of the objects emitting them. The Universe's radio is, by its nature, not that different from visible light – it's electromagnetic radiation, just like light, but at a lower frequency and a longer wavelength so that human eyes can't see it. It is sometimes called the invisible Universe.

This invisible glow was discovered in 1931 by a radio engineer named Karl Jansky, who was studying radio interference in an attempt to improve transatlantic communications. After eliminating the radio noise from thunderstorms (considered the main culprit in disrupting radio signals), he was left with a mysterious static hiss that he eventually calculated had to be coming from the centre of our galaxy, the Milky Way. The refinement of radar in World War II added more evidence that radio waves were arriving from space. Military RADAR (Radio Detection and Ranging) operators detected odd radio emissions that they first thought might be attempts by the Germans to jam their detection equipment. After a few panicked days of observations, they eventually determined that the jamming was coming from the Sun, which was emitting radio waves. They had other more immediate concerns than astronomy, though, and, rather than following up this discovery, maintained their lookout for incoming bombers. It wasn't until after the war that radio astronomy developed as a field of research (in its early days using a lot of now surplus military radar equipment). It quickly became one of the fastest growing and most important elements of astronomy, a science that itself has grown very quickly in the last century. The invisible Universe began to be unveiled.

The chief tool of the radio astronomer is the radio telescope, which is something we're all familiar with even if we don't know it. The humble satellite dish is, in fact, a radio telescope designed to pick up signals from satellites beaming out their own radio "glow." Astronomers' radio telescopes tend to be much larger than a garden-variety satellite dish – sometimes by shockingly large margins. The most famous and recognizable is the monstrous 305-metre Arecibo radio telescope in Puerto Rico, which is built into a natural dish-shaped rocky sinkhole

fault
fault

and has been featured as an exotic locale in Hollywood movies.

Not all radio telescopes are in the form of a dish – some are huge wire webs even larger than the Arecibo telescope. Radio telescopes have to be huge partly because of the size of the radio waves they're trying to capture (which are far larger than tiny light waves), but size also makes them very sensitive and able to capture and focus weak radio waves emitted from across the visible (or should we say invisible) Universe. Large or small, all radio telescopes work in much the same way as optical telescopes. Optical telescopes capture light and focus it on an eyepiece or a camera. A radio telescope captures radio waves and focuses them on a receiver, which works in many ways like a standard radio receiver. It amplifies the signal and can be tuned to listen to specific frequencies, or a range or band of frequencies. (Modern telescopes can handle many frequencies or bands at once.) Fortunately, while radio telescopes need to be carefully engineered they don't need to be as smooth or perfect as the lens or reflector on an optical telescope, so while they need to be big, they don't have to have the extreme engineering perfection of a visible light telescope.

Astronomers using early radio telescopes quickly began to discover some of the most dramatic and energetic phenomena in the Universe, some of them at huge distances from us. For many reasons, these objects couldn't be seen in visible light, sometimes because they were faint in visible light (but very bright in radio waves), sometimes because they were obscured by dust or gas that intercepted visible light but allowed radio waves to pass through. In any case when early radio telescopes were pointed at the sky astronomers saw a very different picture than they'd ever seen through an optical telescope.

One of the most dramatic objects discovered by radio astronomy is the quasar. A quasar is a "QUASI stellAR" object, so named because when astronomers first began seeing them they seemed similar to (but not exactly like) the stars in our galaxy. They were far more distant than any star in our galaxy – in fact, far more distant than many nearby galaxies – but weren't galaxies themselves because, while they had a powerful radio glow, the emission was from a single point and not from the diffuse glow that would indicate millions of stars. Whatever they were, they were among the brightest single objects in the Universe. We now know much more about these quasars, and many have been found at distances up to 13 billion light-years away. The consensus now is that quasars are the glowing disks around black holes that are hundreds of millions of times more massive than our sun. As these black holes attract and swallow gas, debris, and other stars, the material is accelerated by the black hole's gravity and smashes into the rest of the material the black hole is eating. This releases fantastic amounts of energy in radio waves and (we now know) in visible light as well. These super-massive black holes are found at the centre of galaxies, and they can be fifty times brighter than the galaxy around them. The quasar is the most powerful radio station in the Universe.

Nearly as exotic a radio source as the quasar is the pulsar, which broadcasts an intermittent radio signal – a staccato beat across the Universe. These were first observed in 1967 by a team at Cambridge University who were looking for quasars using a relatively crude four-acre radio telescope. What they found was a short radio bleep occurring every 1.3 seconds. Initially the team thought this could have been a man-made radio signal – perhaps a radar beam reflecting off the Moon. When it was determined that the signal was indeed coming from deep space, the next possibility they had to rule out was that it wasn't from an alien civilization attempting to communicate. In fact, what they'd discovered was what has become known as a pulsar. A pulsar, we now think, is a neutron star – the husk of a medium-sized star after it has gone supernova. Pulsars contain up to about three times the mass of our sun, collapsed into an incredibly dense ball of neutrons around 30 kilometres in diameter. Neutron stars rotate at incredible speed, as fast as once per second, and their intense magnetic fields send a beam of radio waves from their north and south poles, shining like radio searchlights across the Universe. If the star has a little wobble in its rotation (as is often the case) these radio beams sweep across the Universe. The pulsar the Cambridge group discovered has a beam that sweeps across the Earth every 1.3 seconds, as the star rotates, allowing its brief signal to be captured by the radio telescope. Since that first discovery, more than seven hundred pulsars have been identified. The radio program they're carrying isn't all that interesting, but the stations broadcasting it certainly are.

Quasars and pulsars are among the most distinctive radio sources in the Universe. However, as radio astronomy has developed, almost everything has come under its lens – or perhaps its antenna. Distant radio-emitting galaxies have been studied. These are entire galaxies emitting a strong radio glow, powered by super-massive black holes similar to quasars. Radio astronomy has provided insight into such mysterious phenomena as dark matter and the Cosmic Microwave Background, which are fundamental to our understanding of the make-up of the Universe and its origins. Closer to home, radio astronomy has been used to study our own sun, the planets, and even comets. Nearly anything that glows in light also glows in radio, and so radio astronomy provided one more way to look at it.

The answer to the question "What's on the Universe's radio stations?" is simple, then. Everything is.

Why is most of the Universe missing?

Astronomers are smart people. They are also, very likely, good at their jobs. So it's a measure of just how awkward and difficult a place the Universe is that they're now convinced that they haven't found as much as 90 per cent of it. They don't know exactly what it is that's missing, or where it might be hiding. This is the problem of missing mass, now more frequently known as dark matter, and it's one of the most intriguing questions in science today.

Dark matter, in many ways, seems more like an astronomical conspiracy theory than anything else. We don't have any evidence for it, but we know it's having a massive influence on large and important things. We don't really know what it is, or what's behind it or how it's doing what it does. We've never seen it, and by its very nature it's going to be practically impossible to detect – almost like it's hiding from us – and yet we know it's out there. Paging Oliver Stone . . .

The theory of dark matter is supported by several kinds of evidence. The first and most important observation is that galaxies behave as if they were much heavier than they seem to be. To reach for another Hollywood analogy, comedy buffs may recall the movie *Shallow Hal*. In it actor Jack Black's character is hypnotized into seeing his elephantine girlfriend as the svelte Gwyneth Paltrow. He has no explanation for the fact that she bowls over passers-by with a slight bump or that chairs collapse underneath her. He can see the influence of her mass, but can't see the mass itself. The situation for astronomers is much the same. When they look into the Universe they see galaxies, or rather the stars that form galaxies. The mass of all of the stars in a galaxy, however, isn't sufficient to explain the gravitational pull the galaxy is exerting on its own stars and on the other galaxies around it.

As far as we know, stars represent most of the normal matter in any galaxy. Our sun, for example, is thought to represent 99 per cent of the mass of our solar system. You would think, then, that if you added up the mass of all the stars in a galaxy you'd get a close estimate of the total mass of the galaxy. Unfortunately for that simple theory, astronomers have seen galaxies in which the outer stars are spinning around the core so fast that there's no way the gravitational pull exerted by the galaxy could hold them together. They should be flying off like four-year-olds on an out-of-control merry-go-round. They've also seen galaxies tugging other galaxies around the Universe in ways that are as inexplicable as a jockey juggling a sumo wrestler. Clearly galaxies are much heavier than their visible mass can explain.

Credit for discovering the problem – which in science is almost as honourable an accomplishment as discovering the solution – goes to two of the giants of modern astronomy, Fritz Zwicky and Vera Rubin. Zwicky first discovered the untoward influence of galaxies on each other in the 1930s, but the result was largely ignored for forty years. Vera Rubin discovered the anomalous speed of stars in the outer parts of galaxies in the 1970s and built up an iron-clad case by observing the effect in more than sixty galaxies, forcing the astronomy community to admit that they had a problem: galaxies have to be at least ten times as massive as they seem to be. More recently, astronomers using the Canada-France-Hawaii telescope, and in a separate project the Chandra X-ray telescope, have detected the influence of dark matter through a phenomenon called gravitational lensing. Light from distant galaxies is being bent by the gravity of nearby galaxies as it passes by them on its way to us. Again, the amount of bending or lensing is far greater than is explained by the observable mass of these galaxies.

The problem is compounded not just by the fact that there has to be extra mass in these galaxies, but what the extra mass has to be: invisible, and almost entirely undetectable. Dark matter is more than just dark. If it were just a huge amount of sooty black stuff – some kind of interstellar dust – then it would be detectable, sort of, because it would block the light from behind it. We'd see it as a kind of gauzy black curtain over the Universe. But there's no dimming or obscuring of the light coming from all parts of the Universe that there would have to be if there was a diffuse cloud of normal matter around every galaxy. So dark matter has to be transparent because we can see right through it. All we do know about dark matter is where it is. For the most part it's in the same general vicinity as regular matter – clumped up in and around galaxies (as opposed to being smeared throughout intergalactic space).

When the problem of dark matter first presented itself, there were two competing theories of what it might be, a debate that rather engagingly became known as the battle of the WIMPS and the MACHOS. MACHOS are Massive Compact Halo Objects. The MACHO theory was that every galaxy was surrounded by a cloud or halo of dark, massive objects like burned-out stars, black holes, starless planets, and smaller things like asteroids. Because these objects would be relatively small and scattered compared to the vastness of space, they would be effectively transparent, or at least they wouldn't obscure our view of the Universe. There would have to be a lot of them, however. Astronomers have been searching for these MACHOS in the halo of our galaxy, the Milky Way, and they've found some (by looking to see if their gravity bends starlight as it passes by them – gravitational lensing again), but at this point it doesn't look as if there are anything like enough MACHOS to account for the huge mass of dark matter.

The current leading candidates for the dark matter are WIMPS, Weakly Interacting Massive Particles. WIMPS, if they exist, are weird. Ordinary matter, the kind that makes up stars, planets, humans, and such, is composed of the familiar building blocks of protons, neutrons, and electrons. WIMPS would be something else entirely – something physicists call "exotic" matter – and we don't actually have a theory to explain precisely what that would be. WIMPS would have some interesting characteristics, though. First of all these dark matter particles would ignore light entirely. All electromagnetic radiation, including visible light, would pass by and through them as if they weren't there. WIMPS in turn would largely pass through normal matter pretty much as if it wasn't there. As you read this, if WIMPS do exist, they are passing through you at a tremendous rate. They can do this because your atoms are mostly empty space, in the same way solar systems are mostly empty space. It's only the electrical charge on the electron and the proton that binds them together that gives them the illusion of solidity to other electrically charged atoms. Dark matter particles have no electrical charge or interaction and can pass through atoms without interference. The other aspect of this strangeness is that dark matter doesn't stick together. Because of its electrical interaction, regular matter clumps together, forming dust, rocks, asteroids, planets, and stars. Not so for dark matter. Gravity would act on it, to keep it in the vicinity of other concentrations of mass – galaxies, for example – but each dark matter particle would be otherwise something of a free agent.

Given all this, you can understand why detection of a dark matter particle is difficult. There are, however, several efforts underway to do so. One of the current efforts is at the Sudbury Neutrino Observatory's SNOLAB, which is a special facility 2 kilometres below the surface of the Earth in a working nickel mine in Northern Ontario. Researchers there are hoping to catch a rare head-on collision between a dark matter particle passing through the Earth and the nucleus of an atom. Dark matter particles, as we've said, can generally pass through atoms, which are largely empty space. They do, however, have mass and substance, and on rare occasions a dark matter particle should hit the tiny nucleus of an atom head on, creating a pretty good bang and releasing energy in the form of a tiny flash. The researchers are hoping that one of their several different designs of supersensitive detector will allow them to catch one of these rare occurrences – the fleeting flash of a dark matter particle bouncing off a nucleus. They have to do this at the bottom of a mine shaft to eliminate the possibility that their detector would see a collision with a normal particle – an energetic cosmic ray, for example – which is much more likely closer to the surface.

Physics can be a strange science. In order to find 90 per cent of the mass of the Universe we're carefully watching the activity of single atoms in the bottom of a mine shaft. Even stranger, though, is the thought that everything we see and everything we're made of represents just a tiny fraction of the Universe. When it comes to what really matters out there – what's really matter, in fact – we're pretty much irrelevant.

Who's stepping on the Universe's gas pedal?

IN THE EARLY 1930S Edwin Hubble discovered that the Universe was expanding. Distant galaxies were flying away from us at fantastic velocities, and so with the passing of time the Universe was getting bigger. That discovery completely revolutionized our understanding of the Universe. What we'd believed was a static and unchanging place was actually an evolving, expanding system.

In 1998, another revolution happened, triggered by a discovery almost as significant as the discovery of the expanding Universe. Astronomers realized that not only was the Universe expanding, its expansion was accelerating. What they saw was about as surprising as throwing a baseball up in the air and watching it accelerate into the sky rather than fall back to Earth. This analogy pretty accurately reflects what they were seeing the stars and galaxies do. Instead of being slowed down in their expansion by the force of gravity, it was as if they are being pushed apart by a mysterious anti-gravity force. The name by which this mystery force has become known is worthy of any fantasy or science fiction epic: it's called dark energy.

The story of dark energy begins, as so many things in modern physics do, with Einstein. When Einstein and his contemporaries were attempting to fit the theory of relativity with the observed behaviour of the Universe they faced a big problem. They were mistaken about a basic aspect of the behaviour of the Universe. At the time, well before Hubble discovered the Universe was expanding, it was thought that the stars hung in the sky in some kind of permanent balance. Newton had recognized the problem with this idea. If gravity operated to draw all matter together, the stars should be falling toward each other. Newton suggested that the infinite Universe was in perfect balance, and every star was equally attracted by every other star, holding everything in place.

Einstein's new theory of general relativity, however, redefined gravity, and in attempting to map it onto the observed Universe, Einstein recognized the same problem Newton did. The static Universe should collapse in upon itself under the force of gravity. Since clearly this hadn't happened, Einstein added a correction in the mathematics of his theory. It was a number, a "fudge factor," that corrected the theory so that the static Universe could continue to exist without collapsing in on itself. The fudge factor became known as the cosmological constant and it was, in effect, a hypothetical push that offset gravity's pull. Einstein had cheated, and knew it, as he famously called this his greatest blunder. The irony is that it may well be a wonderful mistake – an error that turns out to be correct. We just didn't know it for the better part of a century.

When Hubble discovered the Universe was expanding and the theory of the Big Bang was developed, the need for the cosmological constant ostensibly disappeared. The Big Bang was the huge push that was acting to spread the Universe apart. From the moment of the Big Bang, the Universe was coasting on the momentum of that initial push. Gravity should have been acting to slow that expansion over time, just as a baseball thrown into the sky will slow down and reverse course eventually. The great question then for much of the twentieth century was what would happen next? Would gravity eventually win, in time stopping the expansion of the Universe and bringing everything back together again, in a Big Crunch that was precisely the reverse of the Big Bang? This was a popular idea, partly because it provided an answer for the question of where the Universe came from. A Big Crunch would re-create the infinite mass, energy, and density that had existed at the Big Bang, and so another Big Bang could follow. The history of the Universe might just be an infinite sequence of Big Bangs followed by Big Crunches followed by Big Bangs.

In 1998, however, this picture unravelled when Einstein's cosmological constant reared its ugly head again. Two teams of astronomers were using modern telescopes to do what Edwin Hubble had been doing early in the century, looking for mileposts in the distant Universe. Distance is difficult to determine in space since a star or galaxy's apparent brightness can reflect either its size or its distance. A nearby dim galaxy can look the same as a distant bright one. Hubble used a special class of stars called Cepheids to look within our galaxy and at nearby galaxies – ones within several millions of light-years. The nature of these stars allows them to be used as mile-markers in space. Astronomers call these stars, and other objects that can be used the same way, "standard candles."

In 1998 astronomers were doing the same thing with exploding stars – supernovae. The advantage of supernovae is that they're so spectacularly bright that they're visible at vast distances – many billions of light-years, or nearly all the way across the visible Universe. A special class of supernova called a Type 1a always explodes with the same brightness, which makes it a perfect distance marker or standard candle. A Type 1a supernova that is one-quarter as bright as another Type 1a, will be twice as far away (because apparent brightness decreases with the square of the distance). There's another interesting aspect of looking at these distant supernovae. In effect, since their light has taken billions of light-years to get to us, looking at these supernovae is also like looking back in time in the Universe's history. Unfortunately looking at supernovae can be difficult. They are, by their nature, not long-lived. The glow of a supernova lasts perhaps a couple of weeks, and when you're looking for something billions of light-years away, you're also looking for something very small.

By 1998 modern telescopes had made searching for supernovae if not easy, at least possible, and the astronomers working on spotting them had found enough to start to see a pattern developing. When they spotted a supernova they would check for two things: one was its brightness, the other was its red-shift, which was, roughly speaking, a measure of the speed at which it was receding from us and also corresponded to distance. This is because of the Doppler effect, which shifts a receding supernova's light toward the red part of the spectrum. What these two measures told them was that these stars were simply too far away. Gravity, after all, should have been slowing down the rate of expansion of the Universe over time, as it gradually chipped away at the momentum given to everything in the Big Bang. The distance to these most distant supernovae, however, showed that the Universe wasn't slowing down. Over time, in fact, the Universe's expansion has been speeding up. It was very much as if that baseball thrown into the sky had started accelerating away like a rocket instead of dropping back to Earth.

The results were double-checked, and further observations have confirmed them. What this meant to astronomers was that there was indeed some force that has been pushing the Universe apart – fighting, and winning, a battle with gravity ever since the Big Bang. Einstein's great blunder, the cosmological constant, was real, and it must be some kind of dark energy, because we can't see it, except in its effects on the Universe.

Dark energy is a terrible (or wonderful – depending on your perspective) problem for physics. We don't know where it comes from, how it's generated, or precisely how it operates. It seems to be a property of space itself as, strangely, it's increased in power over the history of the Universe's expansion. As the space between the galaxies gets larger, dark energy seems to get stronger. In this sense, it's precisely the opposite of gravity. Gravity gets stronger the closer things are together. As a result dark energy has no influence on us here in our local region of space, or indeed in our galaxy, or even on the local cluster of galaxies in which the Milky Way resides. Gravity dominates in these environments. In the great voids between the clusters of galaxies, however, dark energy seems to be asserting itself, exerting pressure, expanding space, and by creating more emptiness, becoming ever stronger.

The discovery of dark energy has significant implications for the ultimate fate of the Universe. For now, it seems, it rules out the possibility of a Big Crunch in which gravity draws back all matter and runs the Big Bang in reverse. It seems as if gravity is losing the battle for the fate of the Universe, and dark energy will ultimately prevail, expanding space forever. It also means that in time astronomy itself will become very dull indeed. As the Universe's expansion accelerates, eventually everything but our local group of galaxies will pass out of our view. This will, however, take a very long time and the chances that our species – indeed, even our star – will be around to see it are very slim. We can, in some strange way, take comfort in the fact that we won't see those dark days.

WHY DO WE THINK THERE
WAS A BIG BANG?

THE STORY OF THE BIG BANG begins either 13.7 billion years ago, when at our best guess it actually happened, or in the early twentieth century when astronomers realized that the Universe was not static and unchanging, but in motion and expanding. This meant that the Universe must have had a beginning that could be explained. This idea launched one of the richest and most productive debates in astronomy since Copernicus declared that the Earth wasn't the centre of the Universe. The picture that we've arrived at is that the Big Bang was not so much an explosion, but the initially rapid and continuing expansion of a tiny, hot, dense universe into the Universe as it is today. The Big Bang is a model full of interesting holes and mysteries, which physicists and astronomers are still working to solve. These holes reflect the current limits of our understanding of physics and the nature of the Universe.

The Big Bang theory has one thing in common with all the theories that preceded it back to Aristotle and Ptolemy. It's based on an attempt to explain the observed motion of the stars and planets. The difference, and the reason earlier theories were replaced, was that we became better at observing the stars, seeing their motions, and finding systems to explain those motions that were compatible with our understanding of the rest of the natural world. Thus Copernicus's system replaced Ptolemy's in the sixteenth century, and Kepler's replaced Copernicus's soon afterward. Then came along one of the true giants of physics – Isaac Newton – in the seventeenth century, and our picture was again transformed. Newton's theory of gravity entirely revolutionized our understanding of the motion of the objects in the Universe. In fact, it did even more, it created the first model for cosmology that was mechanically and mathematically consistent with the way ordinary things around us worked. The motion of the stars was explicable in the same way as a stone dropping from hand to ground was explicable. Newton, in some sense, brought the stars down to Earth.

He also created a problem that, in part, the theory of the Big Bang is explicitly designed to solve. Newton's theory of gravity states that every object in the Universe is attracted to every other object, with the strength of that attraction diminishing by the square of the distance (his famous inverse square law). This, however, put the Universe on a knife's edge. If everything was attracted to everything else, then the stars should be attracted to each other. As soon as the Universe was created, things should have started to collapse together into great concentrations of mass. The idea of black holes was some centuries distant, but it was a logical outcome of Newton's cosmology.

What Newton proposed to explain why this hadn't happened was the perfect design and creation of the Universe. Newton was a strong religious believer and saw the hand of God in the arrangement of the motions of the stars and planets. His explanation for why the Universe wasn't collapsing was that it was infinite and perfectly balanced. Every star was placed so that its attraction to every other star created a net gravity of zero. In every direction from every star the gravitation pull was equal, so stars were held precisely in place. This required an infinite universe in every direction and a god who was capable of setting up this perfectly balanced universe. In fact, it meant more than this. God would have to continue to work to keep it balanced, as Newton realized that the movements of objects in the Universe would soon destabilize the delicate arrangement and cause gravitational collapse. Newton was, however, comfortable with the idea of a little divine intercession from time to time to keep the clockwork in motion.

In modern terms, this was the first "steady state" theory of the Universe, so called because it explains the Universe as a static and unchanging form. It was, in its way, quite a beautiful idea. It was also, according to the insights of modern astronomy, quite wrong. The galaxies in the Universe are in motion, and in fact they are flying apart at a tremendous rate. Unfortunately it took another three hundred years before the techniques were developed that allowed astronomers to realize this. By the end of the nineteenth century, astronomy had come a long way in the sophistication and size of telescopes, but there was still much that wasn't understood, as even the existence of galaxies was still a mystery. This was largely because there was no way to look at a distant light in the sky and know how far from us it was. This was one of the keys that would lead astronomers to the idea of the Big Bang.

Some huge steps were taken by astronomers like William Herschel (1738–1822), who realized that the sky wasn't the same in all directions. He began to picture the shape of the Milky Way, and even worked on determining our position within it (without realizing that it was just one galaxy among millions). The great problem he faced was that every star appeared as a pinprick of light, so estimating their size and distance was almost entirely a matter of guesswork. Even in the eighteenth century, it was realized that stars might not all be identical. They could be larger or smaller, dimmer or brighter. As a result, looking at a star, even comparing it to other stars around it, couldn't tell you if it was nearer or farther away. A star that appeared dim could indeed be a dim nearby star. It could also be a massive bright star extremely far away. This meant, for example, that before the twentieth century it wasn't in any way clear what size the Milky Way was, whether it had a limit or boundary, or if other galaxies exist. There certainly wasn't any way to determine whether the stars were in motion, because such motion is imperceptible over long distances.

That changed in the twentieth century with a number of breakthroughs. The first and most important was the recognition that the Universe does contain mile-posts of a sort. There are certain kinds of stars that can be used as reference points for distances in the Milky Way, and even beyond. The first recognized ones were stars known as Cepheid variables. In the first decade of the twentieth century, it was discovered that they brighten and dim over time at a rate that is related to their mass. Two stars that brighten and dim at the same rate can be assumed to be the same mass. If one appears dimmer than the other, then that means it is farther away, and that distance can be calculated by determining how much dimmer it is. All of a sudden astronomers had an excellent yardstick for measuring our galaxy. Within a decade astronomers were able to determine several important things: first, the size of the Milky Way, which turned out to be far larger than had been estimated. Second, it was possible to determine where we were in our galaxy, which was a bit of a letdown, since we turned out to be a minor star system on its outer fringes. Just a few centuries earlier we'd thought we were the centre of the Universe. Determining the distances of stars in the Universe was half the battle in developing the understanding that led to the Big Bang.

The next step was the realization that just about everything in the Universe was moving. This was the discovery that later convinced people of the validity of the theory of the Big Bang. This great insight came from Edwin Hubble, after whom the Hubble telescope is named. In the 1920s Hubble first of all found some incredibly distant Cepheid variable stars, which led to the idea that the Universe was bigger than had been imagined and that there were galaxies outside of our own. The Milky Way galaxy was just one island of stars in a universe containing many such islands.

Hubble's next discovery, announced in 1929, was the one that turned astronomy on its ear. He noticed a subtle difference in the light coming from these different galaxies. It all seemed to be at a slightly longer wavelength – shifted a little toward the red end of the spectrum than it should have been. If light were sound, the more distant galaxies were playing slightly lower notes than the nearer ones. The only explanation for this was that these galaxies were moving away from us – at a tremendous rate. As a result, the light waves they emitted were being stretched toward the red end of the light spectrum. This "red shift" in the light from these galaxies, thus, was a measure of their speed as they flew away from us. The Universe was quite literally flying apart. We were living not in a static universe, but in an expanding one.

The revolutionary implication of this idea was not just that the Universe was changing over time, but that it had a beginning. If everything was flying apart, then back in time there was a point where things were closer together. In fact, there should be a time when all the matter in the Universe was concentrated in a single point. The fact that this point had then expanded, spreading apart to create an incredibly huge universe at tremendous speed, was the image that gave rise to the theory of the Big Bang. Seventy years of scientific debate and theorizing and astronomical observations have confirmed this picture, though there are many mysteries still to solve.

WHAT WAS THE BIG BANG (AND WHAT CAME BEFORE IT)?

WHEN EDWIN HUBBLE DISCOVERED in the late 1920s that the Universe was expanding, developing the theory of the Big Bang quickly became one of the hottest projects in science. Hubble's observation that the stars were speeding apart and his theory that the Universe was expanding demanded an explanation. What it suggested was that at one time the Universe was infinitely tiny, hot, and dense, and it was still expanding as a result of the impetus of the conditions that existed at this remarkable time.

This picture helped explain what Hubble and other astronomers could see as they peered farther and farther out into space. It also created a number of problems because now the Universe had to have a history, and that history had to be explained. There was no way to research how it had come into being because there was no way to get at its history. Physics had no way to start answering such questions as when did the stars form and where did the elements came from.

The notion of the Big Bang, first developed by the Belgian Roman Catholic priest and cosmologist Georges Lemaître in the late 1920s, changed all this. Along with other developing areas of physics, such as the understanding of atomic physics and Einstein's theory of relativity, it helped put into motion an incredibly productive effort to explain how we could have got something from nothing. As a result, in the twentieth century much of the physical history of the Universe was explained. We're now pretty confident that we understand, at least in broad strokes, how the Universe evolved for the greatest part of its history. In fact, we can account pretty well for everything that has happened since the Universe was born. The birth itself, and whatever preceded it, is still an open question. Let's just say there's some work still to do. So without further ado . . .

In the beginning (you know we had to start like this) the Universe seems to have been an infinitely hot, infinitely dense concentration of energy, not that we are entirely sure that the words *beginning*, *hot*, and *dense* have any meaning in this context, as we don't have a working theory of physics to describe how anything behaves in these conditions. Once the Big Bang was underway, however we're on slightly more familiar territory. Space started to exist, and the clock began to tick. Then something unusual and important happened: the Universe blew up.

It didn't blow up in the sense of an explosion that blasts energy and matter outward, but in the sense of a balloon inflating. The Universe expanded exponentially, and it did so very quickly – faster than the speed of light. This idea is known as inflation, and it's become the dominant theory to explain this time in the Universe because it neatly deals with several problems physicists have struggled with. One is how the Universe can be as big as it is (which it couldn't be without this sudden early inflation) and another is how it later evolved, developing the concentrations of mass and energy that became galaxies and stars. These questions are complicated, to say the least, but the theory of inflation solves them, so physicists have become quite fond of it. In any case, the Universe experienced a brief burst of incredible growth in a very short time – far less than a billionth of a second.

Then it ends. The Universe was tiny at this point, only tens of centimetres across. For the next few billionths of a second it grew at a fantastic rate, more slowly than in the burst of inflation, but faster than the speed of light. This might seem a bit confusing: as nothing can travel faster than the speed of light, how could the Universe have expanded faster than the speed of light? The explanation is that expansion is not the same as travel. A clumsy analogy is two airplanes flying in opposite directions at their maximum speed – say 500 kilometres an hour. They're flying apart at 1,000 kilometres an hour, but neither is travelling faster than 500 kilometres an hour. The analogy isn't exactly correct, as there were no objects in the Universe at this point, but space itself was getting larger. We did say this was confusing.

During all of this early expansion there was no ordinary matter in the Universe. It was simply too hot for anything like an atom to exist. Things cooled off, though, as space expanded as there was less pressure constraining the energy of the Big Bang, and less pressure means less heat. As the Universe cooled, the building blocks of ordinary matter began to form. First sub-atomic particles like quarks took shape and then as they cooled they combined into protons and neutrons. At this point, the Universe was only one second old, and things were still pretty hot – about a trillion degrees. It took about four more minutes for things to cool enough that atomic nuclei could form as protons and neutrons come together to form deuterium (heavy hydrogen) and helium. The Universe now is filled with plasma – a hot soup of atomic nuclei and electrons. It stayed like this for a very long time.

Around 400,000 years after the Big Bang, the plasma had cooled enough for electrons to settle into co-existence with protons. What existed was a fog of hydrogen, deuterium, and helium. This was the only normal matter in the Universe. It was plenty hot, at a temperature of about 3,000 degrees. Because these electrons were no longer free, they were not intercepting photons any more, and the Universe became transparent. Before this stage it was impossible for light to travel through the plasma – plasma is opaque. There was, however, not much to see. This was the Universe's dark age. There were no stars to illuminate the Universe.

Eventually, the hot fog condensed into discrete clouds, and these clouds collapsed to form the first stars and galaxies, and when the stars lit up, the Universe was something like what we see in the skies today. These first stars burned fast and hot, and in only a few million years they exploded in massive supernovae. Their ashes formed the next generations of stars, which populated the Universe we live in.

This succinct explanation omits a wealth of detail about the Universe's dark ages and how the first stars formed, one of the hottest areas of astronomy today.

That takes care of the time after the Big Bang. The natural question that comes next is where did the Big Bang come from. This is a different question from what was before the Big Bang. When you think about it, that question doesn't make sense, as time as we understand it came into existence with the Big Bang. So where did the Big Bang come from?

Unfortunately that's not a question for which science has a good answer. Observational astronomy, along with particle physics, theoretical physics, and mathematics, has developed the picture of what happened when the Big Bang banged and time started. We're still reaching for understanding about the infinite energy and density that had to exist as a precondition for the Big Bang. At best, this is the realm of (educated) speculation. So little is known that this section of the book might as well be labelled like medieval maps of unknown territories with a large illustrated inscription: "Here be Dragons."

Theoretical physicists aren't scared of dragons, however. Physicists working on String Theory and Loop Quantum Gravity and Quantum Cosmology and the Grand Unified Theory are attempting to develop physical and mathematical models that would describe the conditions that created the Big Bang. These theories are very difficult for laypeople to understand as in essence they are mathematical, and the metaphors used to describe them – like strings – aren't always helpful.

So for now, at least, the question of what caused the Big Bang is best answered with a shrug – or maybe a polite change of subject. After all, what fun would it be if we knew everything?

Why is the Universe's temperature on my TV?

No, WE'RE NOT TALKING ABOUT an unusual entry in the crawl on the Weather Channel ("Horsehead Nebula: minus 200 Celsius with scattered cosmic rays"). What we're talking about is the snow you see on your TV screen when it's connected to an antenna (no cable or satellite feed please), and tuned to a channel on which there's no broadcast. It's not just your TV set spazzing. A small portion of that snow is from the microwave radiation that pervades the Universe and is its basic temperature. In every part of the Universe where astronomers look (with telescopes that are sensitive to microwaves) they can see a faint microwave glow. It's far too weak to cook your food, as it has a temperature of only 2.7 degrees Kelvin – that's a little colder than minus 270 degrees Celsius. This microwave glow is just a special kind of light – with a very long wavelength – which is the oldest light in the Universe.

This radiation is known as the Cosmic Microwave Background, or CMB, and it's the afterglow of the Big Bang. Pictures of the CMB from various telescopes have been called the baby pictures of the Universe and have contributed profoundly to our understanding of how our modern universe, with its concentrations of stars and galaxies, grew from the primordial plasma produced by the Big Bang. The study of the CMB is one of the hottest areas in science and has been responsible for two Nobel prizes so far, in 1978 and 2006.

The first Nobel prize was awarded for the almost entirely serendipitous discovery of the CMB by two young radio astronomers named Arno Penzias and Robert Wilson. In 1964 they were working on rehabilitating a large radio telescope (one also sensitive to microwaves) for use in new studies they wanted to pursue. In calibrating it, they kept getting a weak hum from the instrument. Initially they thought this was a malfunction, and spent fruitless months trying to figure out what was causing it. This included some hard work scraping accumulated pigeon droppings off the instrument that they thought might have been responsible for the noise. Even when the instrument was free of pigeon feces, though, the hum persisted. Ultimately they realized that this hum wasn't a malfunction, but something real, and that it was, in fact, a constant signal from the entire Universe. As it turns out, theoreticians working on the theory of the Big Bang (still controversial at the time) had proposed that exactly this kind of radiation would permeate the Universe. What Wilson and Penzias had discovered was the afterglow of the Universe's initial explosion, and, overnight, the theory of the Big Bang went from, to steal a phrase, outhouse to penthouse. The evidence of the CMB was so strong that only a hard core of Big Bang skeptics denied it. Wilson and Penzias were awarded the Nobel prize in 1978.

The Cosmic Microwave Background, according to theory, was produced by a burst of light that was released about 400,000 years after the Big Bang. The Universe was now cool enough for light to travel through it. Prior to this, the Universe was an opaque, hot, dense stew of photons, atomic nuclei, and free electrons – a plasma not entirely unlike the plasma inside a fluorescent light bulb's tube or the plasma of which the Sun is made. What happens to photons in a plasma is that they get scattered like a flashlight beam trying to penetrate a dense fog. They're constantly hitting free electrons and bouncing off only to hit other electrons. In the Sun, or in a light bulb, some of these photons escape to illuminate us. However in the early Universe there was nowhere to escape to – it was all plasma. So photons bounced around endlessly. It was very bright, but you couldn't see anything. In the words of a song you'll regret having on your mind, the Universe was "blinded by the light."

The transition to transparency occurred when the Universe had expanded enough for the temperature to drop below about 3,000 degrees. The plasma collapsed into normal matter. At that point, the electrons in the Universe could no longer travel freely, and they became bound to the free nuclei floating around, and regular atoms formed. The Universe went from being full of opaque plasma to being full of largely transparent gas. Since the electrons were no longer free to scatter the photons, but tied up in atoms, the universal fog lifted, and the photons in flight at that time continued along on whatever path they had last been set on. It was as if a single flashbulb went off everywhere in the Universe at once, reflecting the shape of the plasma just as it was collapsing – the shape of the Universe, in fact.

This is the oldest light it is possible to see. It reaches us today because at the time it was emitted from every point in the

Universe. When it was generated it would have been an intense light in very short wavelengths. Over time, as it has travelled through the vast distances of the continually expanding Universe, the wavelength of the CMB has stretched out, cooling in a way similar to how gas cools as it expands (that, by the way, is how a refrigerator works). In the slightly more than 13 billion years since the CMB was emitted, it has cooled from 3,000 degrees to just 2.7 degrees above absolute zero. This is the light we detect with our telescopes today.

Among those telescopes is the COBE satellite (for Cosmic Background Explorer) launched in 1989, which returned the first detailed map of the CMB, and which supplied the strongest evidence to date to support the theory of the Big Bang. COBE surveyed the CMB in every direction in the Universe and revealed two vitally important things. First, the COBE astronomers found that when the wavelength of the light from the CMB was plotted on a graph, it made a remarkably smooth curve. This was important, because it confirmed that this radiation was what it was thought to be – the first light in the Universe. If it had been a rougher curve – brighter wavelengths in some places than in others – then it might have been argued that the radiation came not from the Big Bang, but from a large number of single points (ancient stars, for example). When the CMB was found to be a smooth curve, nearly the entire astronomical community let out a deep sigh of relief. In fact, when COBE's first results were announced in the form of a graph presented at a conference, there was a standing ovation. Astronomers really are irrepressible. The Big Bang got a big boost and they wouldn't have to go back to the drawing board to explain the origin of the Universe.

The other finding by COBE was that at the smallest scale there was a little bit of variation across the sky in the CMB. It was almost, but not quite, perfectly distributed. This was a finding that was essential to our understanding of how the early Universe formed and, indeed, we should be very thankful for it. This tiny amount of lumpiness in the CMB made us possible. To be more precise, what it made possible was the formation of galaxies and stars – what astronomers call "structure."

After the Big Bang, the Universe was filled with nothing but a stew of hot particles. As the Universe expanded, these particles moved farther and farther apart. If, however, the density of these particles was absolutely the same all across the Universe (if it was too smooth), then gravity would not be able to pull them together to form clumps that would become clouds of gas, and then stars and galaxies. All the particles would be gravitationally attracted to each other equally, and so they wouldn't clump. The mass of the Universe would have remained a more or less evenly distributed and expanding fog. What COBE revealed was there was an appropriately tiny patchiness to the CMB. This reflected the fact that when the Universe was 400,000 years old, the hot gas of atoms that filled it had tiny variations in density – slightly denser parts, slightly emptier parts. Very slightly, in fact, since the variation was only about one part in one hundred thousand. It was enough that over time gravity could work to amplify the differences, concentrating mass to form gas clouds and then the first stars. Nine billion years later, our sun was born, and 4.5 billion years after that humans came along. So it's these tiny variations in density we have to thank for our current existence. The COBE results, by the way, were what won the second Nobel prize for work on the CMB in 2006.

Work done since COBE's remarkable success has further refined our picture of the CMB. The highest profile one was NASA's WMAP mission, the Wilkinson Microwave Anisotropy Probe, which launched in 2001. It significantly improved on COBE's picture of the Universe, confirming and refining its results.

It's still something of a mystery why there were these tiny irregularities in the density of the primordial gas. The best guess at this point is that they appeared when the Universe was very young. They would have been unimaginably tiny – far smaller than atoms. It's thought that they were caused by quantum fluctuations, which is a messy way of saying essentially random blips in space-time that happen because of the strange rules that govern events at the very smallest scale. These would have grown, through the expansion of the Universe and the passage of time to become the basis for the formation of entire galaxies. So if ever you're wondering what quantum mechanics has to do with your everyday life, now you know. You wouldn't be here if not for its early appearance on the universal scene. Quantum fluctuations: don't start a Universe without them.

HOW WILL THE UNIVERSE END?

WITH THE EMERGING DOMINANCE of the Big Bang theory in the latter part of the twentieth century, the question of how our universe came to be was, to some extent, answered. There was and is still a lot to learn about the details of the Big Bang, but by and large astronomers are satisfied with the theory, and now they can entertain themselves by fleshing it out. Previous theories had suggested that the Universe was eternal – infinite in time and space – and so as time passed we ought not to see any major changes in the way it looked or operated. With the Big Bang, of course, we knew this was not the case. The Universe used to look different and has evolved considerably into its current state. So it makes sense to ask, then, if the Universe had a beginning, will it have an end, and if so, what would it be like?

There is a fundamental difficulty with this kind of study, unfortunately: we can see back in time, but not forward. By looking deeper and deeper into the sky telescopes allow us to see light emitted millions or billions of years ago from stars and galaxies. As a result we can see far into the distance (and the past) to see the Universe's ancient history, and look at closer (more recent) things to see what's been happening since. These glimpses of its history allow us to understand the evolution of the Universe. We can't, however, see the future. There's no way to look into space and see what hasn't happened yet. This introduces certain risks when predicting the future. We can predict how things might work out based on what we know now about the Universe and the physics that govern it. If that knowledge is incorrect or incomplete, however, then those predictions are pretty much useless. They're fun, though, so some scientists just can't resist them. As a result there are three possible futures for the Universe: the Big Crunch, the Big Rip, and Big Chill.

One possibility put forward shortly after the Big Bang theory was the Big Crunch – the Big Bang in reverse. The Big Bang, it's argued, gave the expansion of the Universe momentum, and it's been coasting ever since. Up until the late 1990s it was thought that gravity would eventually overcome the momentum that was driving the expansion of the Universe. The expansion would gradually slow under the influence of gravity, which draws all mass together. At some point, expansion would stop and then reverse. There would be an accelerating collapse as the mass of the Universe became more concentrated. Eventually, at the moment of the Big Crunch, all mass and energy would be drawn together again in a point of infinite density and heat. This, potentially, could spawn another Big Bang, and a brand-new universe would be created. Of course humans couldn't survive this process, but you have to break a few eggs to make an omelette. Universal renewal does have its downside.

The idea of a Big Crunch took a body blow when, in 1998, it was discovered that the Universe's expansion wasn't slowing under the influence of gravity, but was actually accelerating under the influence of the mysterious dark energy. If dark energy persisted in this, then the expansion would continue to accelerate, endlessly pushing the Universe farther and farther apart. Once scientists began to look at different models of the dark energy, however, the picture became less clear. It was possible, for example, that dark energy could evolve over time and that in the future it might work differently. One model suggested that at some time in the future dark energy might reverse itself, becoming an attractive, rather than repulsive force. Then it would start reinforcing gravity, and the Universe might again begin to shrink. If this was the case, the ultimate collapse of the Universe might again be possible. The Big Crunch was back.

There were other possibilities, though, based on different models for dark energy, which suggested other possible fates. In 2003, one of these gave rise to the hypothesis (not quite yet a theory) of the Big Rip. The Big Rip would occur if dark energy ran amok. Dark energy, as it is now understood, seems to be a property of the vacuum. As the Universe has expanded, so has empty space, and thus it's thought that there is more dark energy than there was in the past. It's a theoretical possibility, though, that dark energy could, as space increases, multiply in its power, not just in its amount. It would not just be that there will be more of it, accelerating the Universe's expansion, but that it gains in repulsive strength as well. This would be analogous to the force of gravity increasing in strength, doubling your weight even though the mass of the Earth stayed the same. If this idea is correct, the Universe is in big trouble.

At the moment we think that galaxies, even groups of galaxies, are effectively immune to dark energy. Dark energy has power only over vast empty spaces. Over the comparatively short distances within galaxies, or between clusters of galaxies, gravity is considerably stronger than dark energy, and so keeps things together. If, however, dark energy were to grow in strength, it might become strong enough to overwhelm gravity on these scales as well. It would start to tear clusters of galaxies apart. Over time, as its strength increased, it would tear the galaxies themselves apart, then solar systems, and then even stars and planets. Eventually it might even become powerful enough to tear atoms and sub-atomic particles apart. Shockingly, this would all happen quite soon (at least in cosmological time scales). The Big Rip could happen in only about 35 billion years. In other words, the Universe may already be middle-aged and on its way to an unpleasant and messy end.

Considered more likely, perhaps, than either of these two scenarios at the moment is the Big Chill, otherwise known as the heat death of the Universe. This, unlike the Big Crunch or the Big Rip, is the long slow way to go, and that might be preferable for many. After all, as the old joke goes, "I'd rather die peacefully in my sleep, like my grandfather, than screaming in terror like his passengers."

The Big Chill scenario imagines that dark energy will continue to work much as it does now, accelerating the expansion of the Universe, but having little effect within galaxies or clusters of galaxies. Thus time winds on as the galaxies and stars evolve into the future. What this means in the near term – the next few tens of billions of years – is that the expansion of space will continue, generating more space and increasing the influence of dark energy in the vast emptiness. The space between our galactic neighbour-hood and the rest of the Universe will accelerate and eventually other galaxies will disappear, receding so far from us that their light cannot reach us at all. The Milky Way, then, in a matter of tens of billions of years, will be alone in the visible Universe.

Of course it won't entirely be our galaxy by then. The Milky Way will have merged with the other nearby galaxies, including the Large Magellanic Cloud and the Andromeda galaxy to form a giant elliptical galaxy, more chaotic and less beautifully structured than the spiral arms of the Milky Way. By this time our sun will have burned out into a warm, but no longer burning white dwarf star, and the Earth will have either been consumed by the Sun's end or flung out into interstellar space, a frozen ball of rock and ice. Perhaps, though, we will have found a way to be elsewhere when all this happened, and so might witness what follows.

The Big Chill predicts a long slow decline for the galaxy. New stars are being formed in our galaxy today from the dust and gas left over from the original formation of the galaxy, and also from the debris of giant supernovae. This process may already be slowing down, as much of the free gas from the Big Bang cap-tured by our galaxy when it formed has already been consumed. Supernovae will become less common as well. Most of the stars that form today are smaller than those formed in the past and are not massive enough to explode. Instead, like our sun, they will burn out into cooling cinders, going from white dwarf to black dwarf over perhaps a hundred trillion years. At the same time, on rare occasions, black holes will form, perhaps when neutron stars collide. Black holes already in existence will grow by swal-lowing stars, living and dead. So in a hundred trillion years, give or take a few trillions, all that will be left of our galaxy is black holes and burned-out stars, along with their cold, dead planets.

The black holes will not grow endlessly, though. They will eventually attract much of the matter of the galaxy, but they won't swallow it all. Most of it will just miss them and be flung away into intergalactic space by the slingshot effect of their close calls. This will give them enough speed to reach escape velocity from the shrinking galaxy, and so they will drift away into intergalactic space. In time, there will be nothing left, not even galaxies, as the Universe will exist only as isolated frozen stars and black holes. This will, in the end, take about 10,000 trillion years. The Universe at this time will be very cold, very dark, and very empty. Space itself will continue to expand so that eventually each individual star or black hole will be entirely alone in its own visible Universe.

 · What happens after this is even more speculative. One theory is that over the enormous time scales we're talking about even the ordinary matter making up stars and neutron stars will start to break down. The atoms themselves will degrade through a process called proton decay, so that matter itself slowly evaporates away as a feeble energy travelling through the nothingness. By 10^{38} years (1 followed by 38 zeros) this process will have finished and the only objects left in the Universe will be black holes. Over the long, dark time that follows even the black holes will start to break down, slowly leaking away their mass by emitting something called Hawking radiation. They will have evaporated away by 10^{200} years and the Universe will exist only as empty space, filled with weak radiation that can do nothing but maintain a temperature only infinitesimally greater than absolute zero. That's it, for all eternity: the Big Chill.

So those are our possible fates. Enjoy the Universe – while it lasts.